Signals and Communication Technology

For further volumes:
http://www.springer.com/series/4748

Yifei Yuan

LTE-Advanced Relay Technology and Standardization

Springer

Yifei Yuan
Standards Department
ZTE Inc.
Beijing
People's Republic of China

ISSN 1860-4862
ISBN 978-3-642-29675-8 ISBN 978-3-642-29676-5 (eBook)
DOI 10.1007/978-3-642-29676-5
Springer Heidelberg New York Dordrecht London

Library of Congress Control Number: 2012938204

Printed on acid-free paper

Springer is part of Springer Science+Business Media (www.springer.com)

Preface

Why I Wrote this Book

LTE-Advanced becomes a truly global standard for 4G cellular communications. Relay, as one of the key technologies of LTE-Advanced, can significantly extend the coverage, and improve the system throughput. LTE-A standards and technologies were described in several recent books where the limited pages for relay feature prevent the detailed explanations of the technology. In this book, we tried to provide an in-depth description of LTE-A relay development. More specifically, significant portions are spent on relay channel modeling and potential technologies during the study item phase of the development, although some of those technologies, such as Type 2 cooperative relay, multi-hop relay, relay with backhaul of carrier aggregation, were not standardized in Release 10 LTE. The purpose of those discussions was to offer some insights of relay research in future LTE releases. For Type 1 relay which was standardized in Release 10, our focus is to describe the design principles and rationales of key features, rather than literally explaining the specifications. By doing so, we hope that readers can get the intuitions of major candidate techniques for Release 10 relay, regardless of whether they were adopted in the specifications.

Besides the standardization of relay, some implementation aspects of relay were also discussed with the aim to provide a high-level view on how to build a relay node and deploy the relay systems.

Structure of this Book

The arrangement of the chapters follows naturally the standardization process and implementation steps. It starts with the application scenario and channel modeling, followed by the open study on technology and system performance evaluations, then narrowed down to a short list of techniques that would ultimately be standardized, beginning from physical layer, then upper layer working groups, and then in performance working groups. Once the performance requirements are set, the implementation aspects come next. In the end, we provide the outlook of future relay study.

- Chapter 1: Introduction
- Chapter 2: LTE-A Relay Scenarios and Evaluation Methodology
- Chapter 3: LTE-A Relay Study and Related Technologies
- Chapter 4: Physical Layer Standardization of Release 10 Relay
- Chapter 5: Higher Layer Aspects and RAN4 Performance Aspects
- Chapter 6: Implementation Aspects of Release 10 Relay
- Chapter 7: Outlook of Relay in Future LTE Releases

How to Use this Book

This book is written for researchers and engineers working on wireless communications, in particular, in the field of 3G and 4G cellular communications. Chapters 2 and 3 target for researchers with broader interest in relay and related technologies. Chapters 4–6 would be more useful for engineers specialized in designing and implementing the relay systems. The discussions in Chaps. 1 and 7 are more general and suitable for both researchers and engineers.

Acknowledgments

The author first would like to thank the relay physical layer team at ZTE for their contributions to relay study and specifications in 3GPP RAN1. The team, led by Feng Bi, includes Feng Liang, Shuanshuan Wu, Ming Yuan, Jin Yang, Yunfeng Li, and Xumin Yu. Their original research constitutes a significant portion of this book.

The author also thanks his RAN2/3 colleagues, Mary Chion and Si Chen, and RAN4 colleague Yiqing Cao at ZTE, for their help on some of the chapters in this book.

Relay discussions in this book also touch other key LTE-A technologies such as downlink reference signals, carrier aggregation, heterogeneous networks, etc. He greatly appreciates the valuable suggestions by his RAN1 colleagues Ruyue Li, Wenfeng Zhang, Shupeng Li, Huaming Wu, Zhisong Zuo, Junfeng Zhang in the corresponding sections.

Input from ZTE's product teams is quite appreciated with regard to the implementation aspects of relay.

Finally, the author thanks Professor Hequan Wu, the former vice president of the Chinese Academy of Engineering, for his encouragement in writing this book.

Contents

Abbreviations

ABS	Almost blank subframe
ACIR	Adjacent channel interference ratio
ACLR	Adjacent channel leakage power ratio
ACS	Adjacent channel selectivity
AoA	Angle of arrival
AoD	Angle of departure
ASA	Angle spread of arrival
ASD	Angle spread of departure
BCH	Broadcast channel
CA	Carrier aggregation
CAPEX	Capital expenditures
CCE	Control channel element
CDF	Cumulative density function
CDM	Code division multiplexing
CIF	Carrier indicator field
CP	Cyclic prefix
CPE	Customer premises equipment
CQI	Channel quality indicator
CRS	Common reference signal
CSG	Closed subscriber group
CSI	Channel state information
CSI-RS	Channel state information reference signal
CoMP	Coordinated multi-point processing
D2D	Device-to-device (communications)
DCI	Downlink control information
DeNB	Donor eNB
DFT	Discrete Fourier transform
DL	Downlink
DMRS	Demodulation reference signal
DRB	Data radio bearer
DS	Delay spread

DVRB	Distributed virtual resource block
eICIC	Enhanced inter-cell interference cancellation
EPS	Evolved packet system
EVM	Error vector magnitude
ePDCCH	Enhanced physical downlink control channel
E-UTRAN	Evolved UMTS terrestrial radio access network
FDD	Frequency division duplex
FDM	Frequency division/domain multiplexing
GPS	Global positioning system
GW	Gateway
HARQ	Hybrid automatic repeat request
HeNB	Home eNB
HetNet	Heterogeneous Networks
IoT	Interference over thermal
ISD	Inter-site distance
ITU	International Telecommunication Union
LNA	Low noise amplifier
LOS	Line of sight (propagation)
LVRB	Localized virtual resource block
MAC	Medium access control
MAP	Medium access protocol
MBSFN	Multicast/broadcast over a single frequency network
MCS	Modulation and coding scheme
MIMO	Multiple-input multiple-output (antennas)
MME	Mobility management entity
MU-MIMO	Multi-user multiple-input multiple-output (antennas)
NLOS	Non line of sight (propagation)
OAM	Operation & Maintenance
OCC	Orthogonal code cover
OFDM	Orthogonal frequency division multiplexing
OPEX	Operational expenditures
PBCH	Physical broadcast channel
PCFICH	Physical control format indicator channel
PDCCH	Physical downlink control channel
PDSCH	Physical downlink shared channel
PHICH	Physical HARQ indicator channel
PMI	Precoding matrix indicator
PSS	Primary synchronization sequence
PUCCH	Physical uplink control channel
PRB	Physical resource block
RAN	Radio access network
RA	Resource allocation
RBG	Resource block group
REG	Resource element group
RE	Resource element

RLC	Radio link control
RN	Relay node
R-PDCCH	Relay physical downlink control channel
RRC	Radio resource control
RRH	Remote radio head
RRM	Radio resource management
RRU	Remote radio unit
RSRP	Reference signal received power
RTT	Round trip time
SC-FDMA	Single carrier frequency division multiple access
SCH	Synchronization channel
SCTP	Simple control transmission protocol
SDD	Subcarrier division duplex
SF	Shadow fading or subframe
SFR	Soft frequency reuse
SINR	Signal to interference and noise ratio
SMa	Suburban macro (cell)
SNR	Signal to noise ratio
SPS	Semi-persistent scheduling
SRB	Signaling radio bearer
SSS	Secondary synchronization sequence
SU-MIMO	Single user multiple-input multiple-output (antennas)
TBS	Transport block size
TDD	Time division duplex
TDM	Time division/domain multiplexing
UL	Uplink
UMa	Urban macro (cell)
UMi	Urban micro (cell)
VoIP	Voice over internet protocol
VRB	Virtual resource block
XPD	Cross-polarization discrimination

Chapter 1
Introduction

Relay has been a hot research topic and triggered numerous papers in academia. Its ultimate performance is still an open problem since the relay system capacity is still unknown. For modern wireless communications systems, relay is regarded as a strong candidate for improving the system performance and the coverage at cell edges. Relay, especially the decode-and-forward relay, is proved to outperform the repeaters.

Macro base stations and the low power nodes such as relay node, remote radio head (RRH) pico node and femto node, constitute a heterogeneous network (HetNet) as illustrated in Fig. 1.1. Compared to the homogeneous network that is solely made up of macro base stations, HetNets have quite different topology and interference scenarios.

From the industry standardization point of view, relay is a rather new topic, with respect to technologies such as multiple antennas. Although its history is relatively short, e.g., less than 4 years, three standards for relay operations have already been specified, two for IEEE WiMAX and one for 3GPP LTE. We in this book focus on the technology and standardization of LTE relay.

1.1 LTE-A Technologies

LTE-Advanced is the enhancement of Release 8/9 LTE. The whole standardization is motivated by meeting the performance requirements set by International Telecommunication Union (ITU) for 4G cellular networks to be deployed globally. Narrowly speaking, LTE-Advanced refers only to Release 10 which is a major release compared to Release 9. The broader meaning of LTE-Advanced would include Release 11. There are five key technologies considered for Release 10 as shown in Fig. 1.2: enhanced MIMO, carrier aggregation, enhanced inter-cell interference coordination (eICIC), relay and coordinated multipoint processing (CoMP). During Release 10 time frame, all of them except CoMP were standardized.

Y. Yuan, *LTE-Advanced Relay Technology and Standardization*,
Signals and Communication Technology, DOI: 10.1007/978-3-642-29676-5_1,

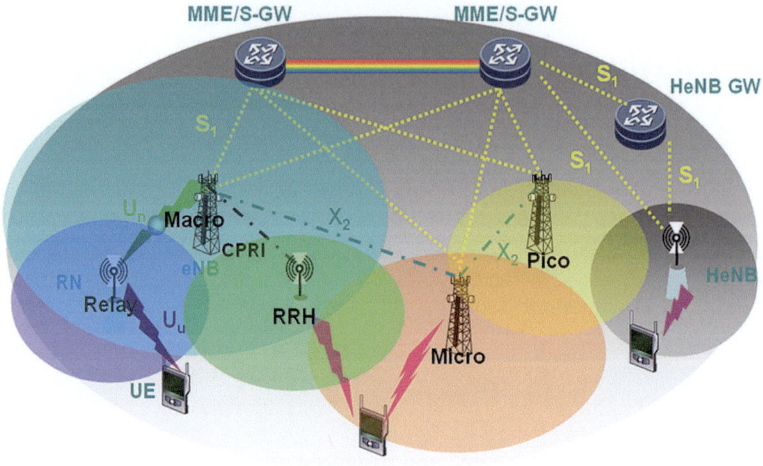

Fig. 1.1 A heterogeneous network with macro, remote radio head (RRH), pico, femto (HeNB) and relay nodes

Fig. 1.2 Key technologies of LTE-advanced in release 10

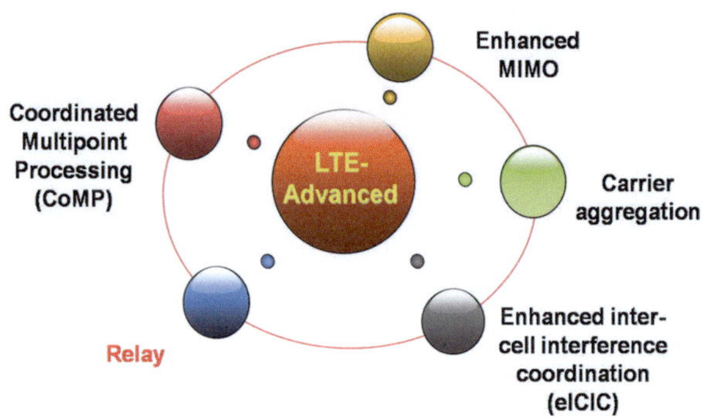

Downlink enhanced MIMO is an extension of Release 9 dual-layer beamforming. Release 9 beamforming primarily works for TDD systems. In Release 10, such beamforming is extended to FDD systems where channel reciprocity generally does not hold. Therefore, channel state information (CSI) feedback is a major specification area. For 8 transmit (Tx) cross-polarization antennas, double codebook structure is specified, which consists of the discrete Fourier Transform (DFT) sub-codebook and the co-phasing sub-codebook, representing the spatial characteristics of the beamforming and the cross-polarization antennas, respectively. The composite codebook is constructed by matrix multiplication of the two sub codebooks, a generalization of Kronecker product of two matrices. The associated downlink signaling, DCI format 2C is defined in Transmission Mode 9, and supports dynamic switching between single-user MIMO (SU-MIMO) and multi-user MIMO (MU-MIMO). A new reference signal channel state information

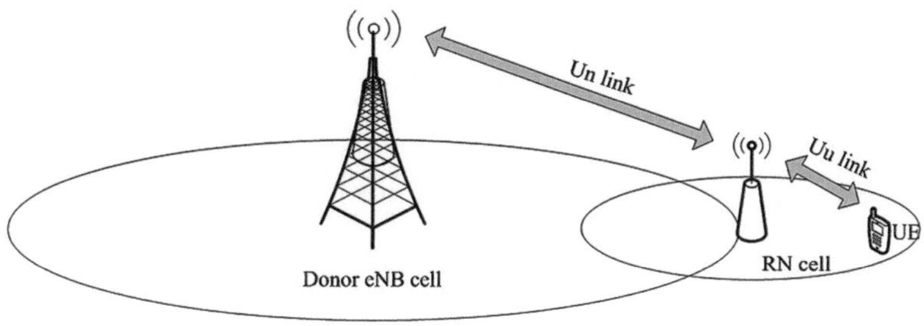

Fig. 1.3 Backhaul (Un) link and access (Uu) link in a two-hop relay

reference signal (CSI-RS), is introduced to facilitate the CSI measurement for SU-MIMO transmission of rank up to 8 and CoMP operations. The configurations for CSI-RS are more flexible than for common reference signal (CRS), both in time and in frequency. In the uplink, the enhanced MIMO supports precoded SU-MIMO of rank up to 4.

From signaling point of view, carrier aggregation allows aggregation up to five component carriers, i.e., 100 MHz, although Release 10 specification optimizes only two-carrier aggregation. In addition to increasing the peak rate, carrier aggregation allows cross-carrier scheduling, which can be used to combat inter-cell interference in heterogeneous networks (HetNets).

In heterogeneous deployment, inter-cell interference can also be mitigated by time domain coordination of transmissions from different nodes, or the so called non-carrier aggregation based ICIC. To achieve that, "almost blank subframe" (ABS) is introduced which contains no downlink traffic channel except some basic signaling channels and reference signals. Almost blank subframes are semi-statically configured by macro base stations and/or low power nodes that have overlapped coverage. The configuration of almost bland subframes should be done in a coordinated manner to minimize the impact of dominant interferers.

Release 10 relay is a decode-forward relay, meaning that the relay node (RN) decodes the data sent from the donor eNB or from UE, re-encodes and forwards to the destination. The relay node has to support LTE Release 8 UEs. The connection between eNB and RN is called backhaul link, or "Un" interface. The connection between RN and UE is called access link, or "Uu" interface, as shown in Fig. 1.3. Backhaul link and access link can be operated in-band or out-band. In-band relay is the focus of the specification. In particular, backhaul link and access link are time division multiplexed in a single frequency band, i.e., only one active at any time.

Relay specified in Release 10 is essentially a stand-alone base station with lower transmit power and wireless backhaul. A Release 10 relay node is typically deployed in fixed locations and shares most of the features of an eNB. It is not designed to exploit the cooperative potential of relaying, a hot topic in academic study. Nevertheless, a lot of experience and lessons were obtained during the Release 10 relay standardization, which will be quite useful for the relay study in

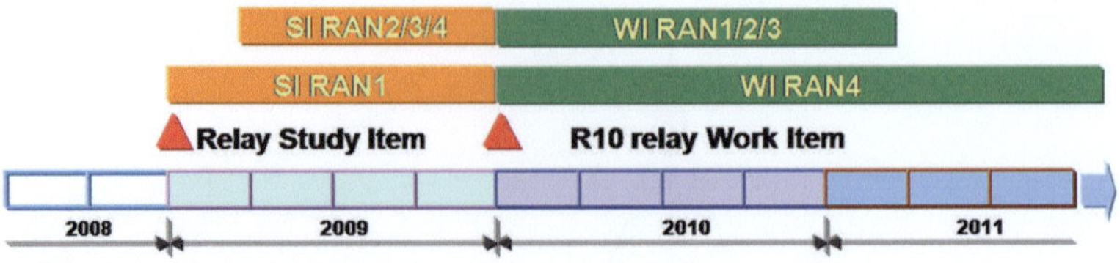

Fig. 1.4 Timeline of release 10 relay standardization

future releases. For example, Type 2 relay, while not specified in Release 10, provides a very good starting point for future relay study, especially for cooperative relays.

Coordinated multi-point processing (CoMP) involves coordinated transmission/reception over multiple points. Those points can be geographically collocated or separated/distributed. When data are available over multiple points, the CoMP transmission is called joint transmission. If data are available only at one point, the CoMP scheme is called coordinated scheduling or coordinated beamforming. The standards development of CoMP was postponed to Release 11 due to some concerns with the backhaul capacity and the spatial channel state information (CSI) feedback for multiple cells.

Major technologies identified and studied during LTE-A Study Item stage were included in 3GPP technical report TR 36.814 [1]. The TR also summarizes the evaluation methodology and simulation results of LTE-A for the submission to ITU-R.

1.2 LTE-A Relay Standardization

In 3GPP, relay study item was officially started in January 2009. The timeline is shown in Fig. 1.4. The study items of other LTE-A technologies also began at that time. By convention, RAN1, responsible for physical layer specification of radio access network (RAN), is the first working group to kick off the study. Slightly later, e.g., in March 2009, RAN2/RAN3, which are responsible for the higher layer and architecture aspects, began their relay study.

In RAN1, the discussions during the relay study item include the application scenarios, categorization of relay, evaluation methodology and high level feature definitions. The relay work progress was slowed down between May and August of 2009 when RAN1 was busy preparing and submitting LTE-A performance evaluations to ITU-R. Since the performance requirements set by ITU-R are for homogeneous networks, relay system performance is not mandatory for the ITU-R submission. Nevertheless, by August 2009, two major relay types, Type 1 relay and Type 2 relay, were identified, and basic parameters for relay evaluation methodology were also agreed. Further refinement of simulation parameters continued till the end of 2009.

Relay work item was kicked off in January 2010, at the same time as the other work items of Release 10. Release 10 relay was primarily for coverage extension scenario, and Type 1 relay was to be specified. At physical layer, the key aspects for relay standardization are backhaul control channels and backhaul subframe configurations. The backhaul control channel specification focused on relay physical downlink control channel (R-PDCCH). The backhaul subframe configurations are coupled with hybrid automatic repeat request (HARQ) timing. The specification consider both FDD and TDD deployments, and in some cases requiring different designs. At higher layers, relay architecture was specified, together with C-plane and U-plane basic procedures. Relay specification work in RAN1, RAN2 and RAN3 was finished in March 2011, slightly later other Release 10 work items. Relay work in RAN4 was not finished in the time frame of Release 10 and the remaining work is continued in Release 11.

1.3 IEEE Relay Standards

IEEE 802.16j includes the most comprehensive relay categories, compared with IEEE 802.16 m and 3GPP LTE-A. In 802.16j, both non-transparent relay and transparent relay are supported. The definitions of "non-transparent" and "transparent" are similar to those for LTE-A relays. A non-transparent relay node has the distinct Cell ID, similar to Type 1 relay in LTE-A. It has the authority of managing the resources and can generate cell control messages. A transparent relay node shares Cell ID and cell control messages with its serving base station, similar to Type 2 relay in LTE-A. 802.16j can support multi-hop relaying, i.e., >2 and mobile relaying.

802.16 m relay focuses only on a subset of relay categories defined in 802.16j. The relay node in 802.16 m is a fixed two-hop non-transparent relay with distributed scheduling, similar to Type 1 relay in LTE.

The radio frame is divided into access and relay zones, used for the transmissions for access link and backhaul links, respectively. The access zone position always precedes the relay zone position. To accommodate the asymmetric DL and UL traffic load, the durations of access zone and relay zone could be different in DL and UL directions. From the aspect of frame structure, there is no particular restriction to support relay zone in 802.16 m. This is a major difference from LTE Release 10 relay where the backhaul subframe allocation is restricted to MBSFN subframes. 802.16 standards support asynchronous HARQ in the uplink. Therefore, 802.16 m relay has more freedom in allocating uplink backhaul subframes, compared to LTE-A relay.

Base station transmits additional MIMO midambles in downlink relay zone for an operational relay node to perform synchronization with the base station. Due to co-existence of mobile terminal and relay node in relay zone, the advanced MAP (A-MAP) control channel and the data channel of access zone are reused in relay zone. The associated channels, e.g. fast feedback control channel, UL hybrid automatic repeat request (HARQ), and sounding channel, should adjust their locations or avoid transmitting on OFDM symbols occupied by switching gaps in relay zone. Overall, the standards impact is smaller than in LTE Release 10 relay.

In 802.16 m relay, HARQ is performed hop-by-hop independently, and so is the header compression. The data plane backhaul control signal interface can terminate either at the base station or the relay node.

802.16 m mobile terminal is aware of relay node including the possible enhanced features, whereas LTE Release 8 UEs are not allowed being enhanced to be "RN-aware". Hence, the backhaul link communication for a LTE-A relay is less flexible than the similar backhaul link of an 802.16 m relay. This disadvantage would be more pronounced in the TDD system due to the further constraints of HARQ timing.

1.4 Book Objectives and Outline

In 3GPP, to minimize the impact on the existing physical layer specifications for eNB-UE connection, a separate specification TS 36.216 [2] was designated for type 1 relay physical layer features. The structure of TS 36.216 is very compact, with the assumption that readers are familiar with Release 8/9 LTE physical layer specifications and can cross reference the corresponding chapters in TS 36.211 (channel structure and format), TS 36.212 (coding and modulation) and TS 36.213 (UE procedures). The very succinct wording in TS 36.216 may require significant amount of effort to fully comprehend the relay operations. Thus, the first objective of this book is to facilitate the understanding of design principles of Release 10 relay and ease the reading of TS 36.216.

The second objective is to provide a comprehensive picture of major relay techniques investigated during the study item stage and the work item stage. These discussions target for readers who have broader interest in relay and related technologies, not limited to the specifications.

As mentioned above, a key specification aspect of Release 10 relay is the backhaul control channel, i.e., relay PDCCH (R-PDCCH). As Release 10 relay node cannot receive PDCCH from donor eNB, certain PDSCH region has to be allocated for R-PDCCH transmission. This prompts the possibility of frequency domain multiplexing (FDM) and use of Release 10 demodulation reference signal (DM-RS) for R-PDCCH. The discussion/standardization of R-PDCCH has a far reaching effect on Release 11 enhanced PDCCH (ePDCCH). From this prospective, the detailed description of R-PDCCH can shed lights on ePDCCH design.

During Release 10 specification, relay timing was one important topic since relay node is half-duplex and needs certain guard period to switch from transmission to reception, or vice versa. To accommodate the guard period and different backhaul propagation delays, several schemes for relay frame timing can be used. Half-duplex operation may become more popular in some emerging technologies such as device-to-device communications. In this sense, relay timing study would provide valuable experience for technologies in future LTE releases.

Type 2 relay, which features cooperation between macro eNB and relay nodes, was studied during the study item. Although not specified in Release 10, type 2 relay showed some promise. The on-going Release 11 CoMP specification, especially Scenario 4, shares some key features with cooperative relay. In this sense, the discussion of type 2 relay in this book is helpful for the understanding of CoMP.

Besides the standards development, implementation is another important step for the relay technology realization in wireless industry. It often requires much more effort and investment for real-world optimizations and developments. One chapter of this book is dedicated to the high level description of relay implementation and deployment.

In academic study, relay is still a topic attracting lots of research interest. Some schemes have strong potentials in improving the system performance and user experience, and thus promising for future wireless systems or standards. The relay study is increasingly connected with other emerging technologies such as device-to-device communications, soft cell network, etc. We use one chapter to provide our thought on the outlook of relay.

The book is organized as follows. Chapter 2 discusses the application scenarios and simulation methodology for relay. Various relay technologies are described in Chap. 3, with performance studies. Chapter 4 contains detailed descriptions of Release 10 relay specifications at physical layer features. High layer aspects and performance requirement are outlined in Chap. 5. Chapter 6 discusses some aspects for Release 10 relay implementation. Outlook of relay and related technologies in future LTE releases are provided in Chap. 7.

References

1. 3GPP TR 36.814: Evolved Universal Terrestrial Radio Access (E-UTRA): Further advancements for E-UTRA physical layer aspects. http://www.3gpp.org/
2. 3GPP TS 36.216: Evolved Universal Terrestrial Radio Access (E-UTRA): Physical layer for relaying operation. http://www.3gpp.org/

Chapter 2
LTE-A Relay Scenarios and Evaluation Methodology

LTE-A relay study item started in January 2009 and ended in December 2009. The study item began with discussions of potential deployment scenarios. Relay scenario identification provided valuable guidance for the discussion of evaluation methodology. Through proper channel modeling of the target scenarios, we can accurately assess the relay system performance, which is crucial to the technical choices and business judgment of each scenario. Considering that evaluation methodology is a specialized area that is not directly tied to the relay technology itself, we use a separate chapter discussing the evaluation parameters, with the focus on channel modeling.

2.1 Relay Scenarios

A number of potential deployment scenarios are of interest to major operators [1]. Although not all of them were prioritized for Release 10 relay specification, the discussion is helpful for scenario identification of future relay and related technologies.

2.1.1 Rural Area

As Fig. 2.1 shows, the rural service features wide coverage area and low user density. The low user density leads to rather uniform and thin distribution of users where ubiquitous coverage becomes crucial. So the first question of interest to the operators is how to lower the deployment cost. In this sense, relay would provide an efficient solution in reducing the number of macro eNBs. As the environment is primarily thermal noise limited, the issue of weak signal affects not only data

Y. Yuan, *LTE-Advanced Relay Technology and Standardization*,
Signals and Communication Technology, DOI: 10.1007/978-3-642-29676-5_2,

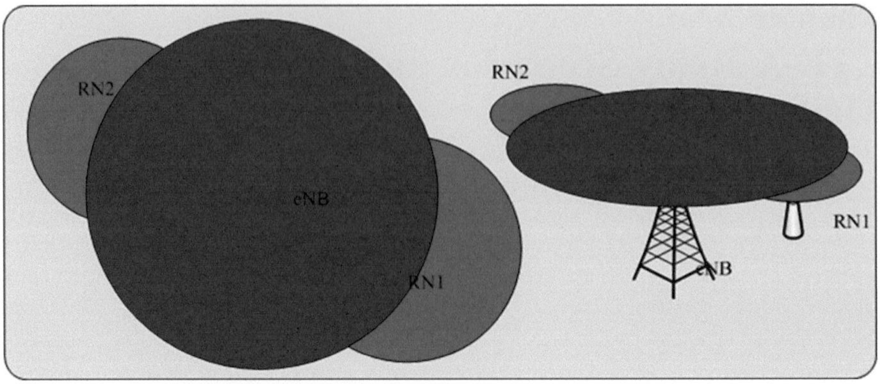

Fig. 2.1 Rural area scenario for relay

channels, but also the control channels. That is, UE within the coverage of a relay node would not be able to decode correctly the Layer 1 control signaling from eNB. Given the limited number of UEs served by each relay node (RN), the coverage, rather than the capacity improvement per RN cell would be the major concern.

The long distance between eNB and RN means low signal to noise ratio (SNR) at RN receiver. The situation prompts the need for decode and forward relay to improve the SNR at the UEs served by RNs. Amplify and forward repeater is not suitable here as the noise at RN receiver is also amplified by the repeater, i.e., no SNR improvement.

To reach out more UEs without deploying too many relay nodes, the transmit power of RN can be relatively high, and the coverage of each RN can be several kilometers. The actual coverage depends on the operating band, and the propagation environment. In general, due to the low–rise morphology, the line of sight (LOS) propagation would be dominant. NLOS is relatively rare unless the terrain is very hilly and/or covered with tall vegetations.

The high transmit power deployment favors fixed location of relay node, and the site planning is very crucial.

2.1.2 Urban Hot Spot

Urban hot spot is just opposite to rural area scenario. The user density is quite high and often non-uniformly distributed, as illustrated in Fig. 2.2. The main objective is to enhance the capacity. Therefore, the coverage of each RN is relatively small and many RNs could be deployed within a macro eNB coverage area. There could be a lot of coverage overlaps between RNs. Due to the densely deployed macro eNBs and RNs, interference scenarios become very complex and difficult to predict. HARQ is crucial to ensure reliable transmissions. The challenging interference environment makes conventional repeater unsuitable in this scenario.

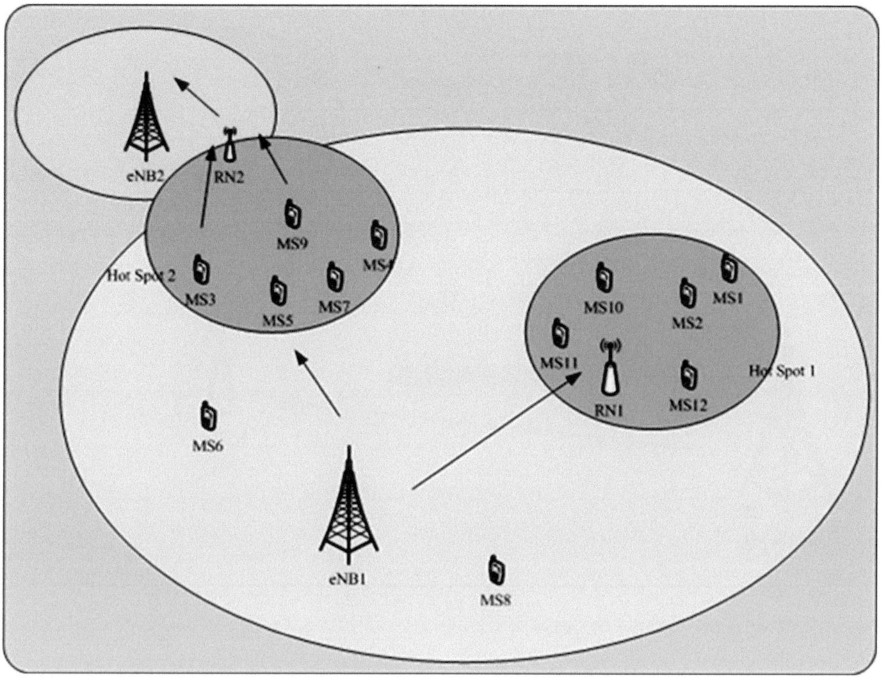

Fig. 2.2 Urban hot spot scenario for relay

Either fixed location or nomadic relay node can be deployed to alleviate the zoning regulation and renting cost for the installation. Transmit power of RN tends to be low, so that (1) the interference to neighboring cells is small; (2) RN can be made compact to allow more flexible deployment.

High–rise building in urban area results in strong non-line-of-sight (NLOS) propagation environment. Channel modeling, both for the backhaul and the access link, needs to capture dominant NLOS and the shadow fading. Note that in this scenario, the tall buildings are not yet big enough to completely block the macro eNB's signals to create coverage holes.

2.1.3 Dead Spot

In the concrete-jungle like urban terrain, quite often the height of macro eNB antennas is significantly lower than the nearby buildings, as Fig. 2.3 shows. The immerse size of surrounding high–rises can easily create dead spots in their shadows. The propagation environment may be similar to the urban hot spot, with the exception that shadow fading can have larger standard deviations. Significant building blocking creates an isolated area where signals from nearby eNBs can

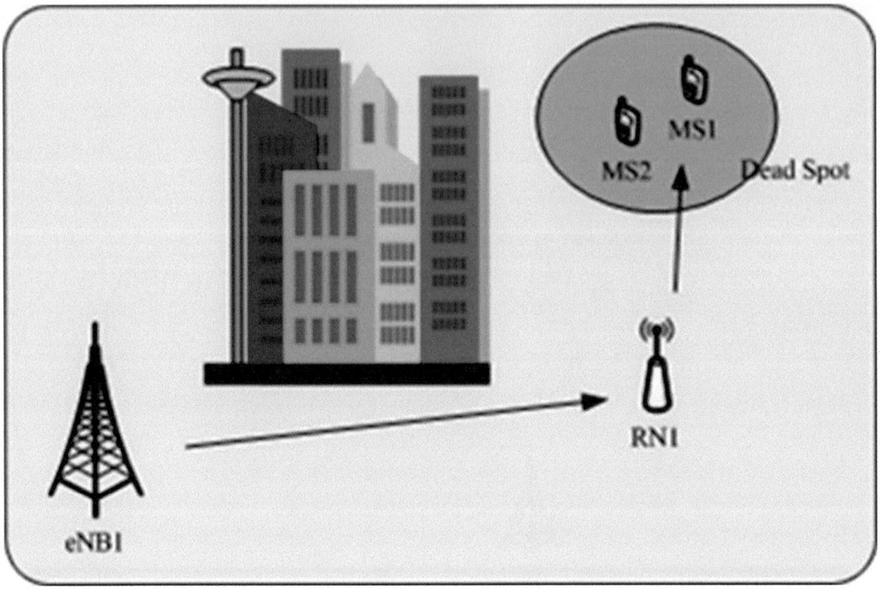

Fig. 2.3 Dead spot scenario for relay

barely reach. The interference is mainly the thermal noise where conventional repeater could be an alternative solution, in addition to the decode and forward relay.

2.1.4 Indoor Hot Spot

Relay could be used to achieve high data throughput for indoor hot spot, as Fig. 2.4 shows. This scenario is different from the urban hot spot in the sense that majority of users are indoors and stationary. The shadow fading tends to be high due to the wave reflection and refraction against the walls. The relay is supposed to provide enhanced throughputs and to serve indoor users in low coverage areas (e.g. deep indoor, or in buildings far from the donor eNB), similarly to a home eNB (femto cell). Offloading from the donor eNB may also be possible since the backhaul link is with higher quality compared to the direct link and thus requires less resource from the macro cell than the indoor UEs would. Therefore, the macro cell capacity is increased.

In the absence of wired backhaul, the indoor relay would be particularly important in the following environments:

- Far-apart houses in suburban and rural areas. Outdoor relays may not be feasible due to the extremely large number of outdoor sites needed to achieve enough good coverage.

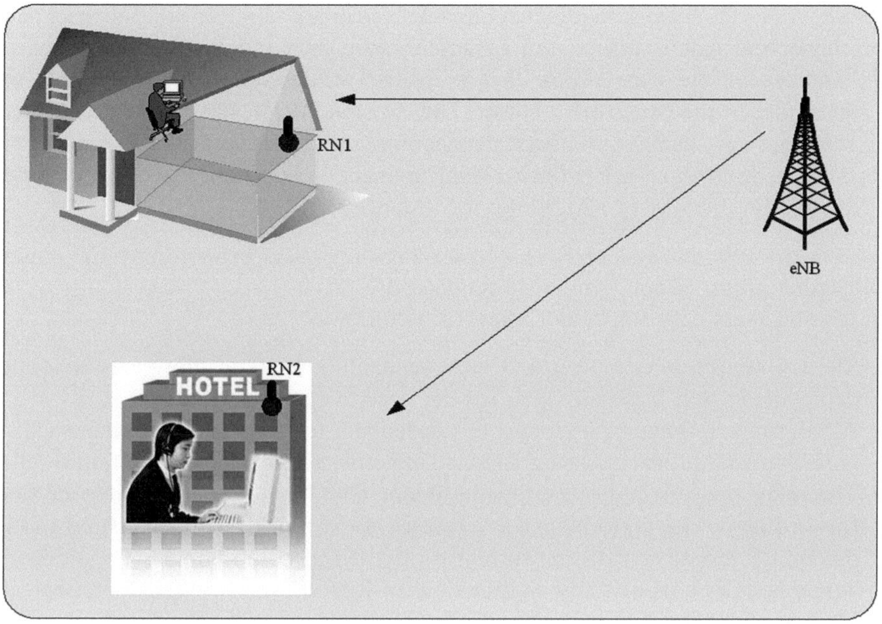

Fig. 2.4 Indoor hot spot scenario for relay

- Within high–rise buildings, users at different floors would experience vastly different channel qualities, for example, more LOS is expected at high floors, so that the signal from the serving eNB and the interference from neighboring eNBs are both strong. Thus the signal to noise and interference (SINR) would be quite poor. In this case, a directional antenna on RN pointing toward the donor eNB can effectively enhance the signal strength from the donor eNB, and suppress the interference from neighboring eNB, thus improving the SINR of the backhaul.

From the aspect of the use case, indoor relays bear a lot of similarities to femto cells. Two usages have been identified [2]:

- Residential usage: the RN serves a block of apartments or a house where the number of UEs is small, i.e., <4
- Business usage: the indoor relay serves a floor of an office building, or a shopping mall. It is expected that a large number of UEs (typically ranging from 30 to 100) would be supported by the RN

The backhaul link may suffer building penetration loss if the relay backhaul antenna is inside the building. To ensure that RN can offer better performance compared to the macro and UE connection, the RN's antenna of the backhaul link should be placed in a location that would result in good backhaul connection. A few installation methods can be considered:

1. The RN can have two distinct modules: a donor module for the backhaul connection that is placed, for example, close to a window, and a coverage module for the access link that is placed where coverage is needed, for example, in the center of a house. The two modules could be connected in a wireless way, using an outband connection (e.g. unlicensed 5 GHz band).
2. The donor antenna of RN for backhaul connection is installed above the clutter height, on the roof of a building, for instance.

It is possible that indoor RNs would serve only users belonging to the closed subscriber group (CSG), similar to femto cells.

The business case for indoor relays is as follows:

- The low transmission power and self-backhauling features allow very low-cost deployment; Self backhauling saves the need for cable installation.
- ADSL subscription may no longer be needed, nor for other cable services such as optic fibres. This provides a significant competitive edge compared to femto cells.
- The relay node is controlled by the donor eNB and operates in decode and forward mode. Its performance is expected to be much better compared to the indoor L1 repeaters that are currently installed at some homes. The network would be fully aware of any malfunctions of RNs without special Operation and Maintenance (OAM).

In this scenario, it is likely that the relay node would be self-installed as a customer premises equipment (CPE), rather than planned by the operators. Consequently, it is difficult to perform site optimization to improve the channel quality of backhaul link.

2.1.5 Group Mobility

As the penetration rate of mobile phones especially smart phones keeps increasing, users on public transportations would have more propensity to access high speed wireless services. Voice services on buses or trains are typically cacophony and the data rate is low. Battery life is shortened, in particular to overcome the penetration loss through the vehicles and the high Doppler. The battery power drain is also due to the continuous measurement carried out by on-board UEs, in both idle and active mode to accommodate the frequent handover caused by the group mobility. In such condition, UEs would experience excessive rate of broken connections since the mass number of on-board UEs frequently trigger the simultaneous handover and cause serious signaling congestions, leading to higher call-drop rate.

In such case, a relay node can be deployed on the roof-top of the moving train or bus as Fig. 2.5 shows, to serve on board passengers, and to alleviate the problem of vehicle penetration loss. The challenging link seems to be the backhaul which suffers fast fading due to the movement of the vehicle. Passengers are supposed to

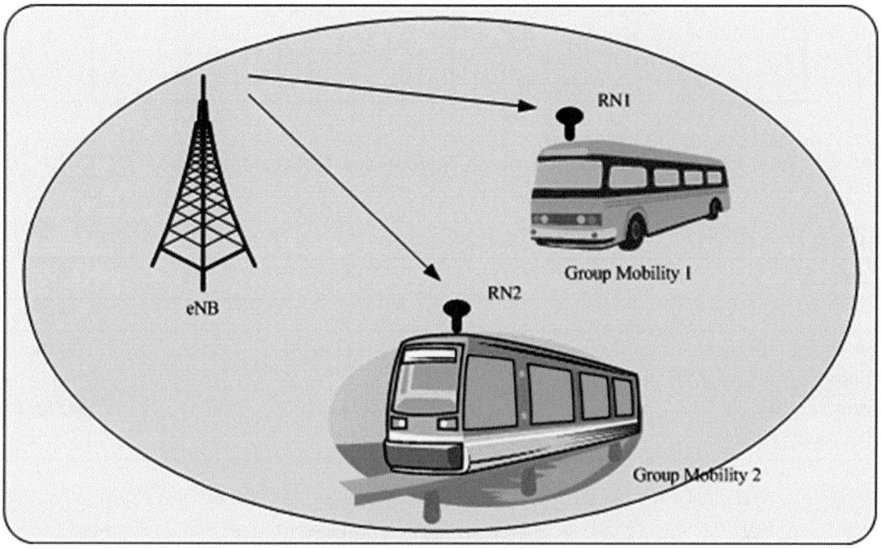

Fig. 2.5 High group mobility scenario for relay

rather stationary relative to the vehicle, so that the access link would be strong and stable.

The key usage of relay in this scenario is to "aggregate" multiple UEs' connections on a vehicle to a single access point. Obviously, it cannot be achieved by conventional repeaters that are totally transparent to on-board UEs.

Recently, relay for high speed trains has gained significant interest. Building high speed rail has become the national key projects in some countries. To provide the high speed communications for on-board passengers is also part of those national-key projects. Fast communications are crucial as passengers on high speed trains are more likely to be data-hunger professionals and would access the internet and emails when on-board. The capacity requirement for relay backhaul is expected to be very high, considering the high density of users on a train. So the demand for high data rate is high for both downlink and uplink traffic.

Backhaul channel characteristics, including pathloss, shadow fading and fast fading, would be different from those of eNB to UE connection, and some of them could be relatively benign, for example,

- Higher elevation of mobile relay antennas mounted typically on top of train roof ($\sim$5 m).
- Terrain and morphology along the rail track tend to have less scatterers, resulting in strong line of sight (LOS) propagation, except under bridges or in the tunnels.

However, the extreme high velocity poses significant challenges for the wireless backhaul transmission, even in LOS environment. A particularly difficult issue

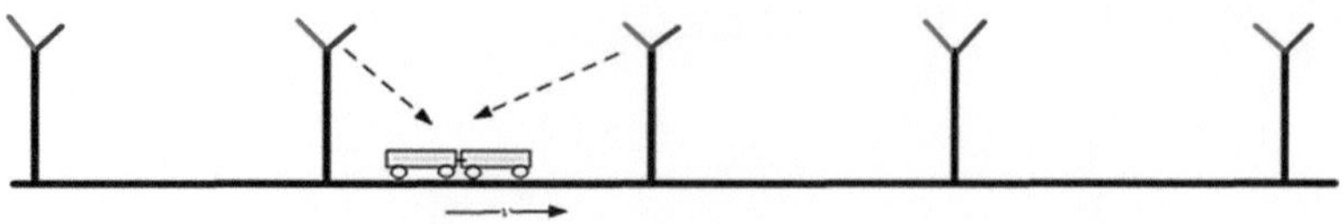

Fig. 2.6 High speed train passing a series of transmission points along the rail

Table 2.1 Parameters for high speed train scenarios

Parameter	Value	
	Scenario 1	Scenario 3
Site-to-site distance	1,000 m	300 m
Minimum track to eNB distance	50 m	2 m
Train velocity	350 km/h	300 km/h
Maximum doppler	1,340 Hz	1,150 Hz

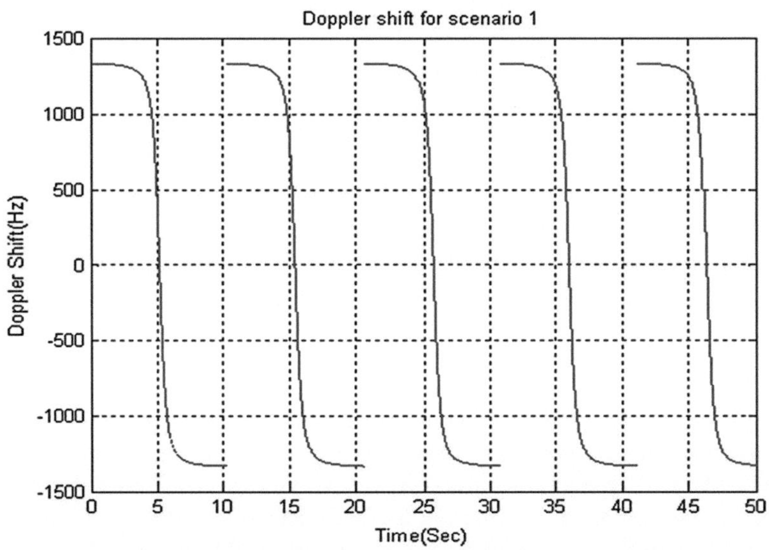

Fig. 2.7 Doppler shift trajectory for Scenario 1 when the track to transmission point distance = 50 m

is the abrupt flip of Doppler when the train passes a transmission point as illustrated in Fig. 2.6. To show how steep the Doppler transition, let us look at an example that uses a simple LOS model [3] to describe the Doppler shift changes.

Two scenarios are considered with parameters listed in Table 2.1. In Scenario 1, the train speed is a little higher than in Scenario 2, thus leading to higher Doppler. The inter distance between adjacent transmission nodes is also longer. However, the rail track in Scenario 1 is further away from the transmit points, compared to

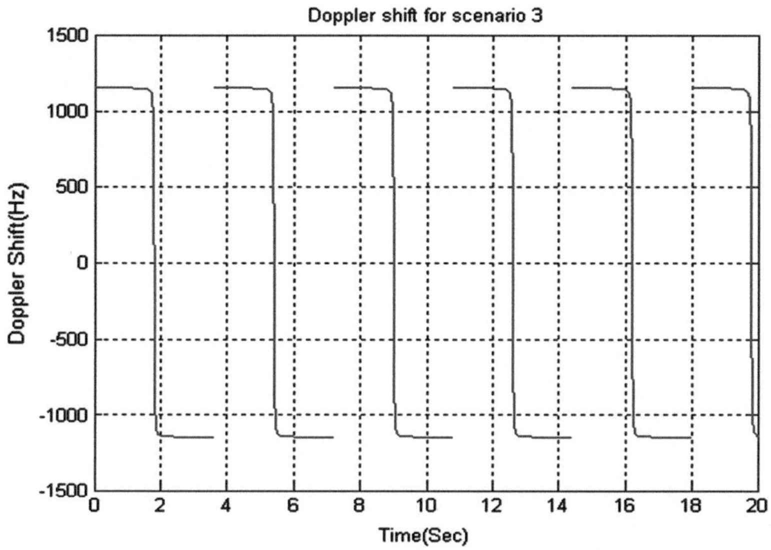

Fig. 2.8 Doppler shift trajectory for Scenario 2 when the track to transmission point distance = 2 m

Scenario 2. The changes in Doppler as a function of time are depicted in Figs. 2.7 and 2.8, respectively. It is seen that the Doppler can flip between the two extremes in a few milliseconds. Note that the transmission point switching is considered in Figs. 2.7 and 2.8.

2.1.6 Emergency or Temporary Network Deployment

The self-backhaul nature of decode and forward relay can be used to provide temporary wireless network in the case of an emergency such as natural disaster seen in Fig. 2.9, and terrorist attack, or during an event such as sport games, public gathering, outdoor concerts, etc. In either case, temporary network needs to be quickly deployed to fulfill at least the partial functionality of a full-blown network.

2.1.7 Wireless Backhaul Only

In certain rural areas, the cost of laying wired backhaul would be prohibitive, yet operators have some vacant spectrum. Therefore, relay can be deployed to act solely as the backhaul link between eNBs, without serving any UEs, as seen in Fig. 2.10. Using the vacant LTE spectrum has the benefit of tolerance against the

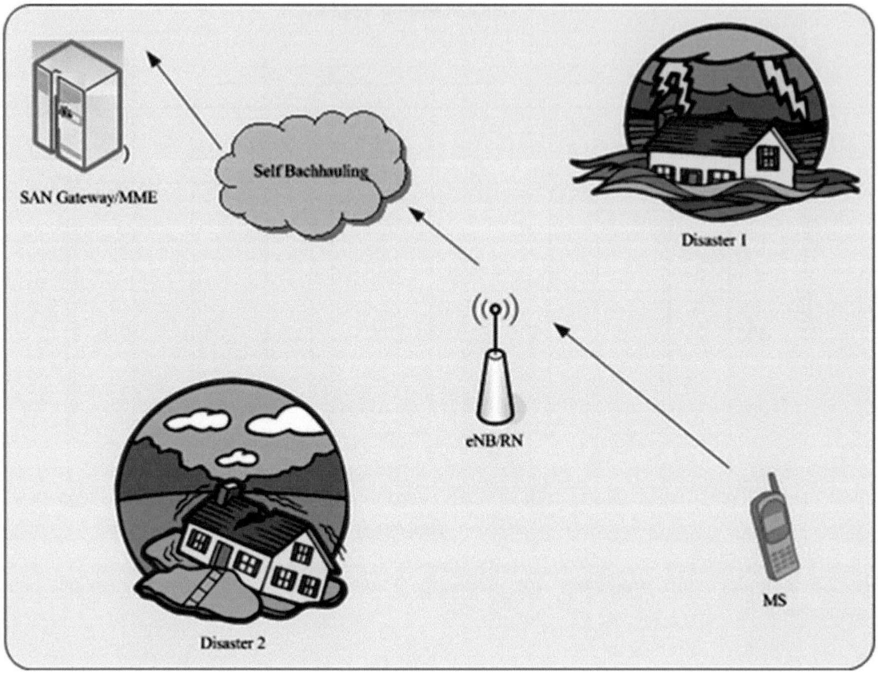

Fig. 2.9 Emergency network scenario for relay

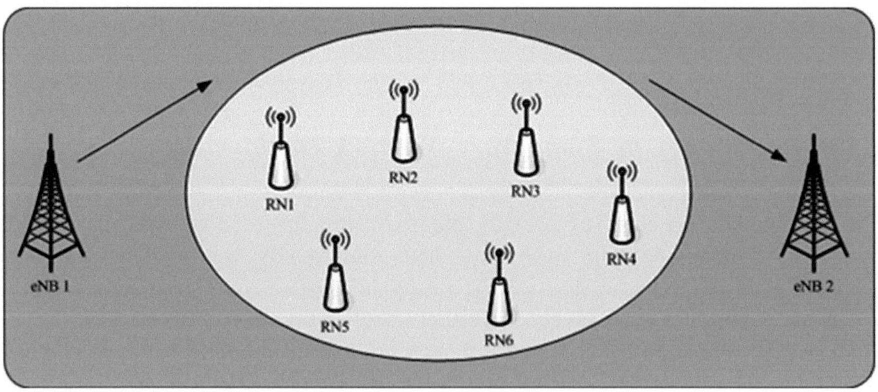

Fig. 2.10 Relay for wireless backhaul only

weather dependent signal attenuation, compared to using shorter wavelength microwave for point-to-point transmission. It would also save the potentially high cost of separate RF module for shorter microwave transceivers.

Relay node in this scenario is with fixed location. Hence, relay site optimization is crucial to ensure the good channel quality of the backhaul link. The transmit power of relay node is expected to be high to guarantee high-speed transmission in the wireless backhaul.

2.2 Channel Modeling

During LTE Release 8 study, a single pathloss model was used for macro eNB to UE connection which is based on the traditional formulae for NLOS propagation environment, with minor correction to account for the contribution of LOS component. That assumption makes sense for homogeneous networks in which the site-to-site distance is constant and the topology of the entire cell grid is regular. However, using single pathloss model may not be accurate enough in heterogeneous deployment as macro eNBs and relay/pico/femto/RRH have quite different transmit powers. The antenna gains, antenna heights and down-tilts are different too. Also, cell topology becomes more diversified in HetNet, which demands more sophisticated channel models to represent the actual propagation environment.

IMT-Advanced channel model is a geometry based stochastic channel model. It was proposed for the evaluations for radio interface technologies. The framework of the primary module is based on WINNER II channel model. It is characterized by the bandwidth of 100 MHz with center frequency between 2 and 6 GHz. Five test scenarios are defined and two or three levels of randomness are introduced where the probability density functions (PDFs) are extracted from extensive measurement data.

The WINNER models are notable in the following aspects.

- The veracity of the models is justified by the extensive measurement data carried out in myriad locations, environments over long time of the observation. The statistical significance of the models is high.
- Several typical propagation environments are defined, and channel models are fine-tuned to each environment. Parameters are environment dependent.
- LOS and NLOS are separately modeled, each with its unique propagation mechanism and parameters. Mixture of two is done statistically, i.e., a UE is instantiated with either LOS or NLOS propagation at the beginning of each run or drop, with certain probability as defined in the model.

Relay is one category of low power nodes of heterogeneous networks (HetNet) which also includes remote radio head (RRH), pico node and femto node. However, their specification areas are quite different. The relay specification in RAN1 deals with wireless backhaul channel optimization, whereas the HetNet in RAN1 study primarily targets for inter-cell interference coordination (ICIC), assuming that the macro node and the low power node operate in the same carrier. For eICIC, the actual work was only started after September 2009, nine months after the relay study item had started. Simulation methodology discussion of relay, in particular for outdoor relay, was later reused for channel modeling of the pico node and RRH.

Channel modeling includes the following aspects:

- Large scale fading which is the most relevant to coverage prediction and interference analysis.
- Delay spread which captures the frequency selectivity of wideband channels
- Angle spread that reflects the spatial richness and ergodic/outage capacity of the MIMO channel.
- Cross-polarization discrimination (XPD) that is important to polarization diversity.

The last three aspects constitute small scale fading. In Release 10, new large scale fading models were proposed for relay backhaul link (eNB–RN) and access link (RN–UE). Those models were based on the real measurements reflecting the typical relay deployment. Large scale fading models were adopted for relay simulation methodology, as well as for pico/RRH in eICIC. Small scale fading models for relay backhaul and access links were proposed, however, due to the limited time, they were not agreed. Companies may use Typical Urban with fixed correlation matrix, or ITU, SCM models or their simplifications for fast fading modeling.

2.2.1 Large Scale Fading Modeling for RN–UE Connection

While WINNER models provide a rich tool for channel modeling, the original work was primarily for homogeneous deployment. WINNER indoor models can be used in some HetNet scenarios such as femto node. However, those models are not suitable for outdoor relays that serve UEs in more general application scenarios. Thus, for relay, pico and RRH, new channel models are needed. Some wireless operators such as China Mobile carried out a series of campaigns on channel measurement for low power node deployment. The vast measurement data makes sure that the derived parameters would be statistical significant. The modeling follows the same procedure as in WINNER project, e.g., LOS and NLOS are treated separately, and parameters are environment scenario dependent.

The path loss models for outdoor node generally follow the form of [4]

$$PL = A\log_{10} d[m] + B + C\log_{10}\left(\frac{f_c[GHz]}{5.0}\right)$$

where d is the distance between the transmitter and the receiver in meters. f_c is the system frequency in GHz. Parameter A, B and C are obtained by curve fitting the measured data.

Let us first look at the channel modeling for access link. Figure 2.11 shows the modeled pathloss in LOS dominant scenario where three curves are compared. The blue curve is obtained by linear fitting of the measured data to average out the perturbations due to the shadow fading. The measurement band is 2.35 GHz, and the height of the outdoor relay node antenna is 5.5 m [4]. Note that those settings are slightly different from the agreed 2.0 GHz operating band and 5 m relay height for system simulations. Some adjustments were applied to the original measured

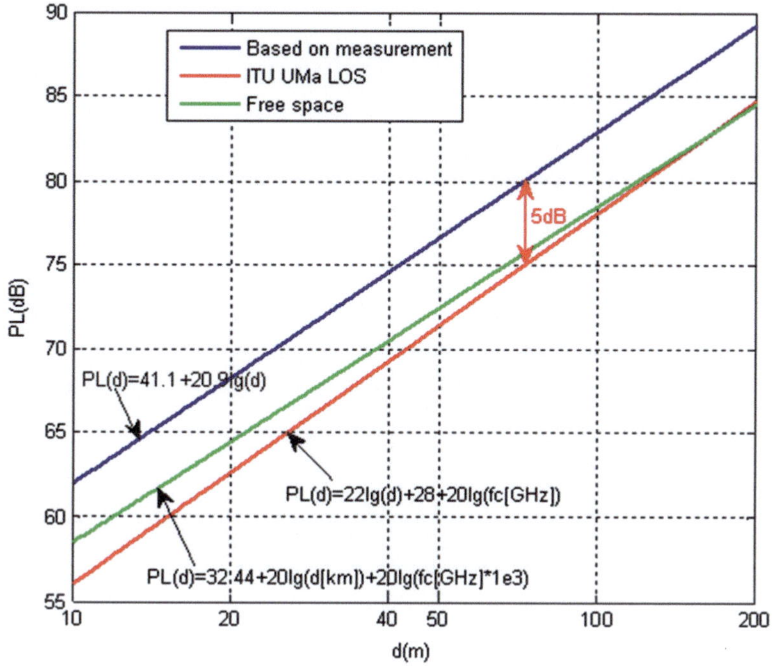

Fig. 2.11 Pathloss comparisons for outdoor relay node to UE connection in LOS dominant environment

data in order to compensate the frequency and antenna-height difference. The curve fitting is done in logarithmic domain, resulting in the following pathloss formula

$$PL(d) = 41.1 + 20.9\log_{10} d[m] \quad \text{or} \quad PL(d) = 103.8 + 20.9\log_{10} d[km]$$

Also plotted in Fig. 2.11 are the pathloss curves of ITU UMa LOS model and free space propagation model. The very close behaviors of these two curves indicate the rather strong free space propagation in urban macro (UMa) LOS model. The slopes of the blue curve based on the measurement and UMa LOS model are quite similar. Their difference is mainly at the vertical interception. On average, the gap is about 5 dB, which can be explained by the lower height of relay node antennas, i.e., 5.5 m, compared to 25 m typically for urban macro cells. That would bring more signal attenuation, due to the higher probability of blocking by ground vegetation, moving vehicles and other obstacles.

Measurement was also carried out for RN to UE connection under NLOS dominant environment [4]. Similar to the measurement of LOS environment, the carrier frequency is 2.35 GHz and the height of the RN antenna is 5.5 m during the measurement. Four straight curves are plotted in Fig. 2.12. The formula based on the measurement data is

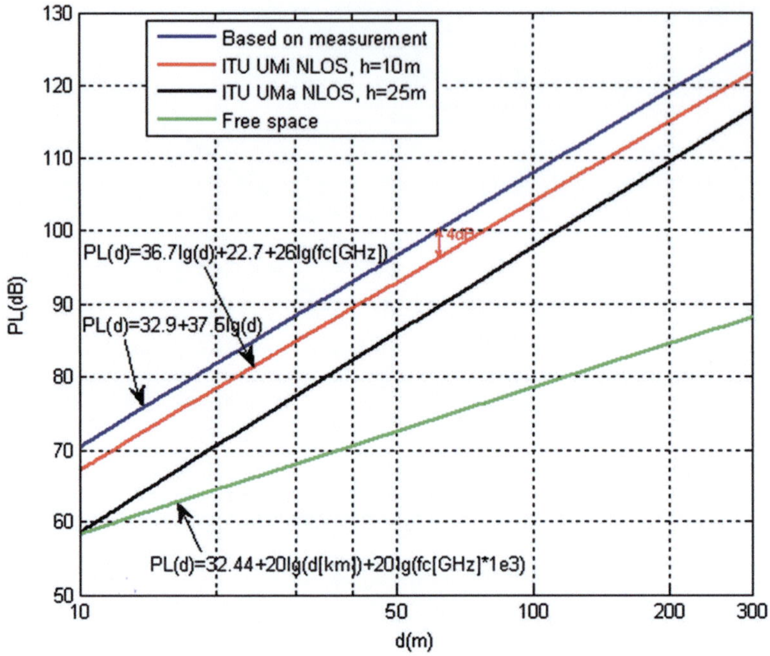

Fig. 2.12 Pathloss comparisons for outdoor relay node to UE connection in NLOS dominant environment

$$PL(d) = 32.9 + 37.5\log_{10} d[m] \quad \text{or} \quad PL(d) = 145.4 + 37.5\log_{10} d[km]$$

The other three curves represent the pathloss in ITU urban micro (UMi) NLOS, ITU UMa NLOS, and free space models. The slopes of NLOS curves are significantly steeper than that of the free space model which is essentially LOS. The three NLOS curves mainly differ in the vertical interceptions. The trend is similar to that in LOS, i.e., as the antenna height is reduced from 25 m in UMa, to 10 m in UMi, and further down to 5.5 m in relay node, the pathloss penalty is widened roughly from 6 to 10 dBs. All these observed from the measurement data reflect the increased probability of obstruction and scattering in the propagation path, as the antenna height is reduced.

2.2.2 LOS Probability of RN–UE Connection

We see from Figs. 2.11, 2.12 that the pathloss in LOS is significantly smaller than in NLOS. For example, at 50 m distance, the pathloss of LOS is about 77 dB, whereas the pathloss of NLOS is 97 dB. The difference is roughly 20 dB. In WINNER model, the propagation environment of a UE can be either LOS or

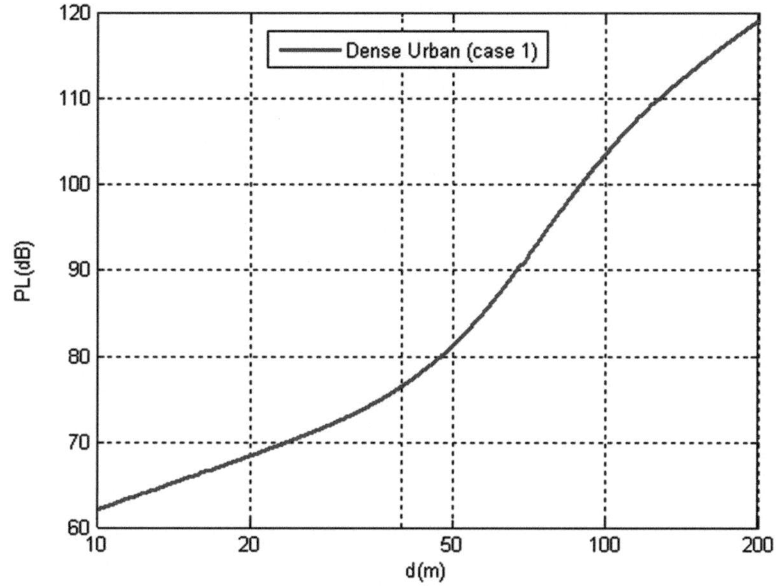

Fig. 2.13 Average pathloss vs. distance of RN–UE, considering LOS probability, Dense Urban

NLOS. The likelihood is governed by the LOS probability which is scenario and distance dependent. In general, the closer a UE is to a macro node, the more likely that the propagation between the UE and the macro is LOS. The LOS probability functions in WINNER can be useful, but they cannot directly be copied to relay scenario.

In light of this, similar formula would be used for RN to UE connection as in the case of ITU UMi, with revised parameters to reflect the smaller coverage for a typical relay node and the lower antenna height compared to ITU UMi. The modified LOS probabilities are as Dense Urban (case 1):

$$Prob(d[m]) = 0.5 - \min\left(0.5, 5\exp\left(-\frac{156}{d}\right)\right) + \min\left(0.5, 5\exp\left(\frac{d}{30}\right)\right)$$

Suburban (case 3):

$$Prob(d[m]) = 0.5 - \min\left(0.5, 3\exp\left(-\frac{300}{d}\right)\right) + \min\left(0.5, 5\exp\left(\frac{d}{95}\right)\right)$$

To get a feeling of the pathloss with LOS probability taken into consideration, we plot the

$$PL(d[m]) = Prob(d) \cdot PL_{Los}(d) + [1 - Prob(d)] \cdot PL_{NLOS}(d)$$

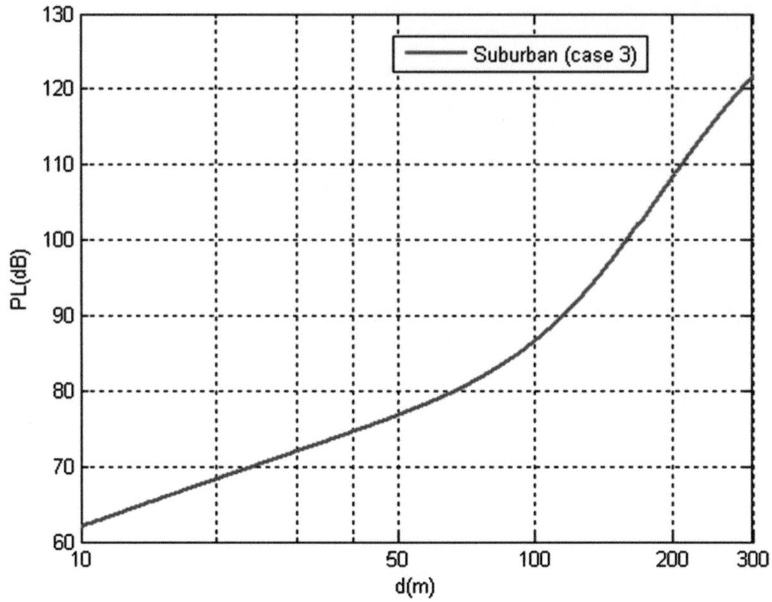

Fig. 2.14 Average pathloss vs. distance of RN–UE, considering LOS probability, Suburban

n Figs. 2.13 and 2.14 for dense urban (Case 1) and suburban (Case 3), respectively. As expected, as the UE is further away from the RN beyond 50 m (for Dense urban) and 120 m (Suburban), pathloss increases rapidly due to the environment shift from LOS to NLOS. Note that such combined pathloss is only in average sense. i.e., the expected pathloss averaged over a number of independent UE droppings. In system simulations, each UE of each drop can either be LOS, or NLOS, not both.

2.2.3 Large Scale Fading Modeling for eNB–RN Connection

We now look at the pathloss model between eNB and outdoor RN. The data were obtained from the same measurement campaign by China Mobile [5], at 2.35 GHz frequency and with 5.5 m antenna height at RN. The blue curve in Fig. 2.15 is obtained by curve fitting of the measurement data in LOS dominant scenario and has the following formula

$$PL(d) = 100.7 + 23.5 \log_{10} d[km]$$

Smaller pathloss is observed in eNB to outdoor RN connection, compared to ITU UMa LOS model and even to free space model at close distance. The reason can be

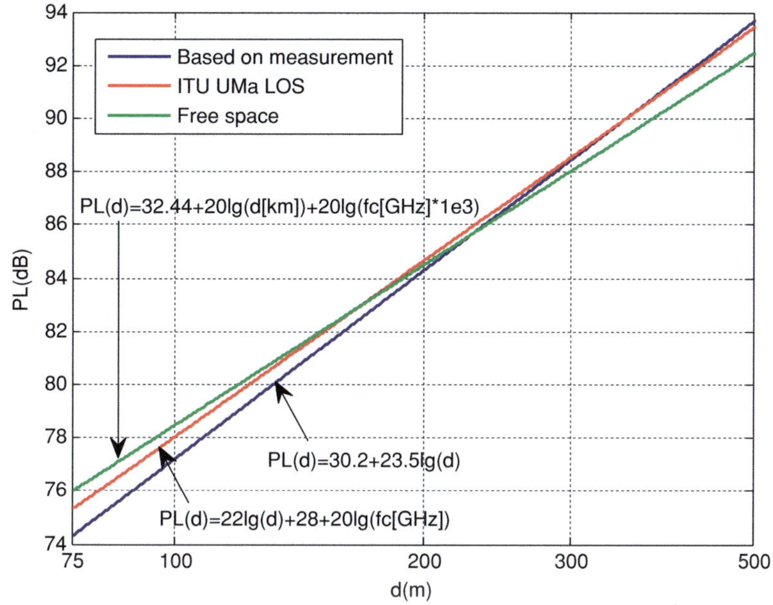

Fig. 2.15 Pathloss of eNB to outdoor relay node connection in LOS dominant environment, compared to other links

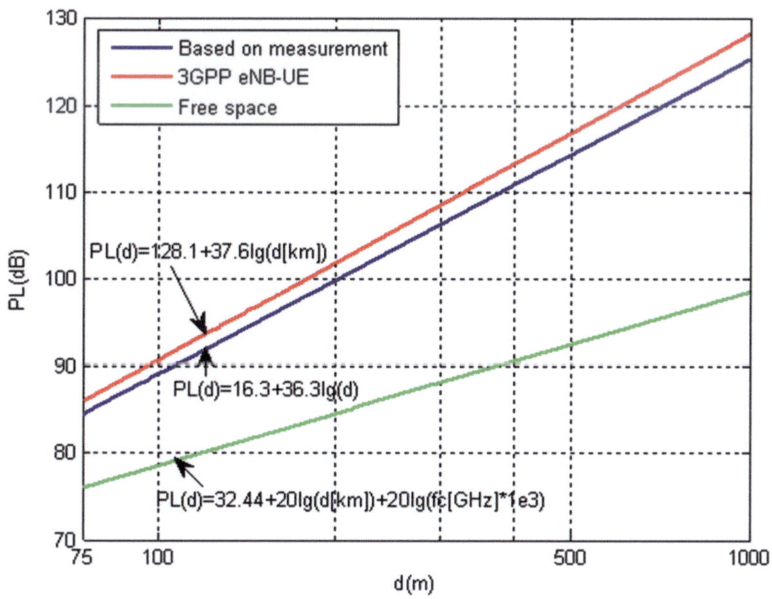

Fig. 2.16 Pathloss of eNB and outdoor relay node connection in NLOS dominant environment, compared to other links

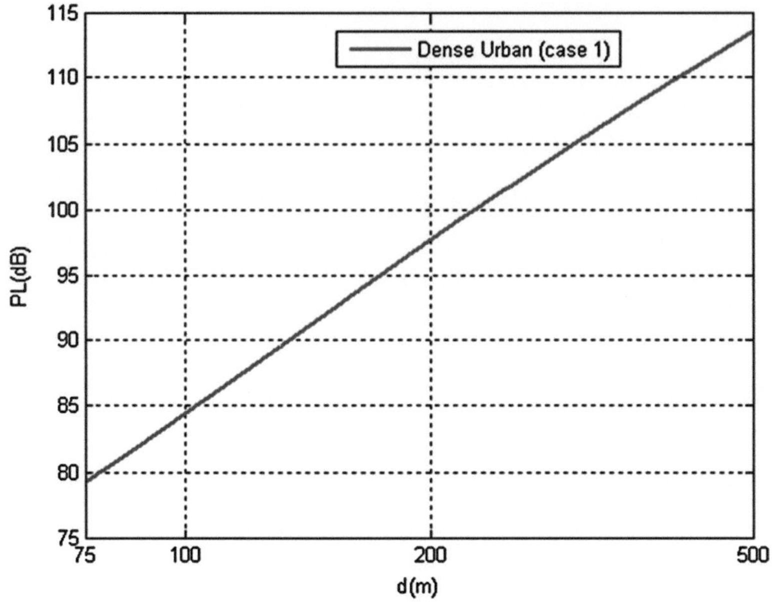

Fig. 2.17 Average pathloss vs. distance of eNB–RN, considering LOS probability, Dense Urban

explained by the higher RN antenna, i.e., 5 m height than that of UE which is assumed to be 1.5 m.

The eNB–RN pathloss in the case of NLOS are compared in Fig. 2.16. The 3GPP eNB–UE model is a NLOS model, widely used for macro cell deployment. The so called "3GPP eNB–UE model" was a model that does not distinguish LOS and NLOS (although NLOS is assumed dominant), i.e., all eNB–UE connections use the same pathloss equation. It had been used till August 2009. For the measurement data, the curve fitting leads to the following pathloss equation for eNB–RN link in NLOS dominant environment.

$$PL(d) = 125.2 + 36.3 \log_{10} d[km]$$

2.2.4 LOS Probability eNB–RN Connection

In dense urban (Case 1) environment, the LOS probability for eNB and RN connection can be based on ITU UMa model, with certain adjustment to account for the higher antenna elevation at RN compared to UE. The LOS probability is expressed as

$$Prob(d[m]) = \min\left(\frac{18}{d}, 1\right)\left(1 - \exp\left(-\frac{d}{72}\right)\right) + \exp\left(-\frac{d}{72}\right)$$

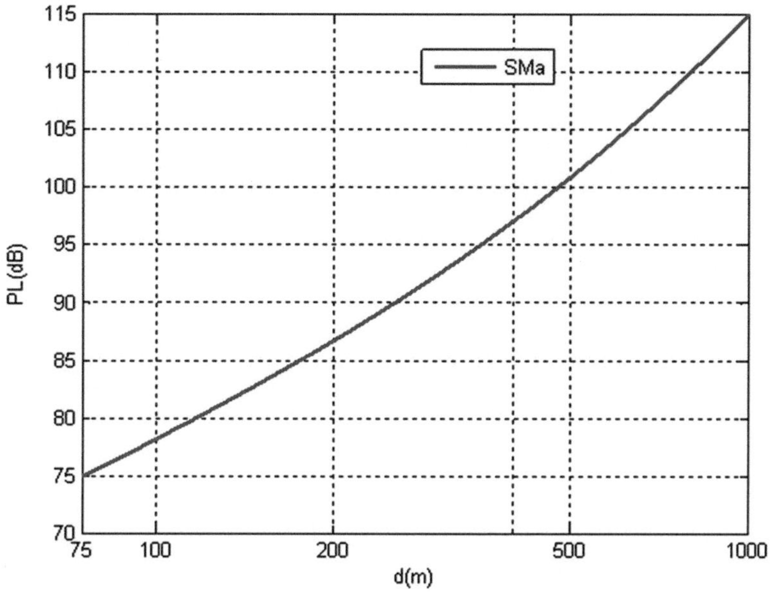

Fig. 2.18 Average pathloss vs. distance of eNB–RN, considering LOS probability, Suburban

Following the same rationale, LOS probability for suburban macro (SMa) can be adjusted to reflect the less attenuation due to the RN antenna height. The formula for suburban environment is

$$Prob(d[m]) = \exp\left(-\frac{d-10}{1150}\right)$$

Similar to the case of RN–UE connection, we plot in Figs. 2.17 and 2.18 the average pathloss as a function of distance, taking into account of LOS probability, for Dense Urban (Case 1) and Suburban (Case 3), respectively.

Note that the LOS probabilities above simply capture the propagation environment as a function of eNB to RN distance. They do not take into account the relay site optimization which would generally improve the probabilities of LOS propagation and be favorable to the backhaul transmissions. The impacts of RN site optimization on channel modeling will be described subsequently.

As discussed in Sect. 2.1.4, the indoor relay scenario is very similar to that of femto cell deployment. In WINNER channel modeling, the elaborate indoor models already include various cases of wave propagations in different floor settings, those channel models were quickly adopted for femto study in eICIC as well as for indoor relay study. Therefore, the discussion of indoor relay channel modeling is skipped in this chapter.

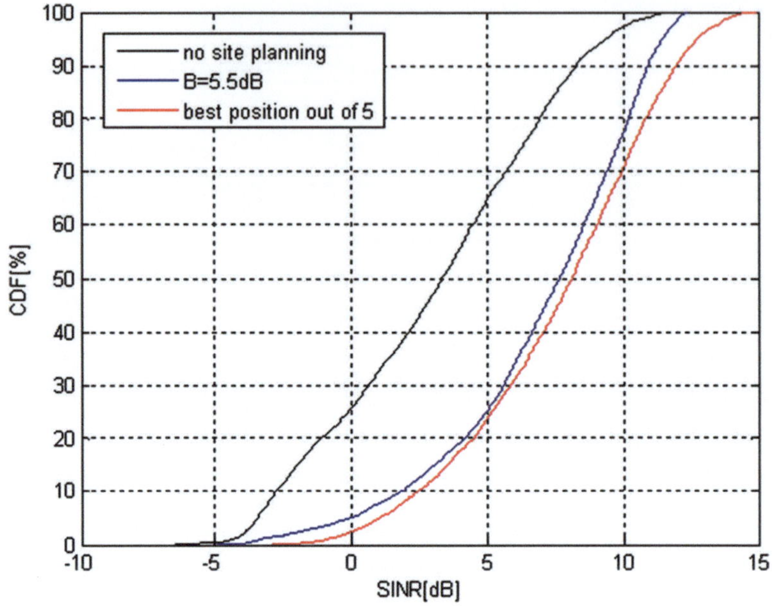

Fig. 2.19 Pathloss bonus reflecting the gain of post-site-planning geometry

2.3 Impacts of Relay Site Planning

For the discussion so far, the favorable pathloss of backhaul link is hinged on the higher antenna at relay node. In the fixed relay deployment, especially for outdoor relays, site optimization can further improve the propagation environment for the wireless backhaul communications. In 3GPP, such improvement comes from the two aspects in channel modeling: less signaling attenuation from the donor eNB, and increased LOS probability of backhaul link.

2.3.1 Less Attenuation from Donor eNB

A simple method is to add pathloss bonus directly to the connection between the eNB and a RN. Note that the bonus only applies to RNs with NLOS propagation with eNB, and only to the link to its donor eNB. In another word, a relay node of LOS propagation with its donor eNB, or of LOS/NLOS propagation with its neighboring eNBs do not enjoy such gain.

One method of choosing optimal bonus value is to compare the RN geometry curves between different bonus values to match the SINR gain due to RN site planning. For example, Fig. 2.19 compares the post-site-planning geometry curve to the RN geometry curve with B = 5.5 dB bonus, but without site planning.

Table 2.2 Site planning bonuses added on pathloss (dB), searching radius of 50 m

ISD (m)	eNB antenna beamwidth	eNB-RN distance ratio of ISD	Bonus (dB)
500	120	0.2	4.9
		0.5	4.4
		0.6	5.2
	60	0.2	2.6
		0.5	3.7
		0.6	4.2
1,732	120	0.2	4
		0.5	4.2
		0.6	4.8
	60	0.2	3.3
		0.5	3.8
		0.6	4.3

Table 2.3 Site planning bonuses added on pathloss (dB), searching radius of 30 m

ISD (m)	eNB antenna beamwidth	eNB-RN distance ratio of ISD	Bonus (dB)
500	120	0.2	3.9
		0.5	3.4
		0.6	3.7
	60	0.2	2.9
		0.5	2.9
		0.6	3.1
1,732	120	0.2	3.1
		0.5	3.3
		0.6	3.6
	60	0.2	3.1
		0.5	3
		0.6	3.1

The inter-site distance (ISD) is 500 m in the simulation. The macro to relay distance is 250 m, i.e., 0.5 ISD. Five candidate relay sites ($N = 5$) are considered within a searching area of 50 m radius around the virtual relay node, i.e., randomly dropped RNs. It is observed that the effect of applying 5.5 dB pathloss bonus on the backhaul link is almost equivalent to that of 5-site optimization within 50 m.

Similar exercises were carried out for different macro inter-site distances, eNB to RN distances and eNB antenna patterns [6]. The fitted bonuses are summarized in Tables 2.2 and 2.3. In some sense, site planning takes advantage of shadow fading that is random. However, there is certain distance dependent correlation btween the candidate sites. The correlation follows a circular exponential decay function. Therefore, the smaller the search area, the lower the gain of site planning. Such expectation is confirmed by the comparison between Table 2.2 of 50 m radius and Table 2.3 of 25 m radius. Similarly, wider radiation pattern, i.e., 120°, helps to include more potential relay nodes, so that site planning would bring more gains. The bous is not very senstivie to the eNB to RN distance, nor to the macro inter-site distance.

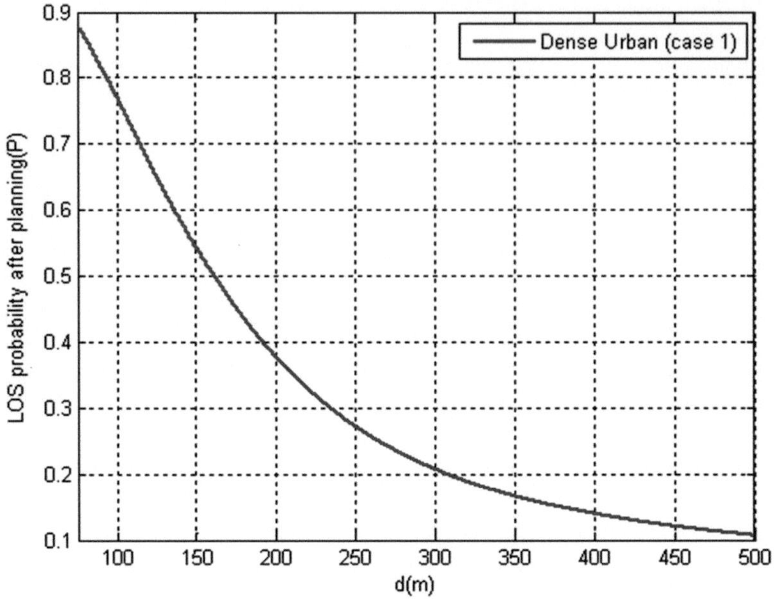

Fig. 2.20 Post site planning LOS probability vs. distance of donor eNB–RN, Dense Urban

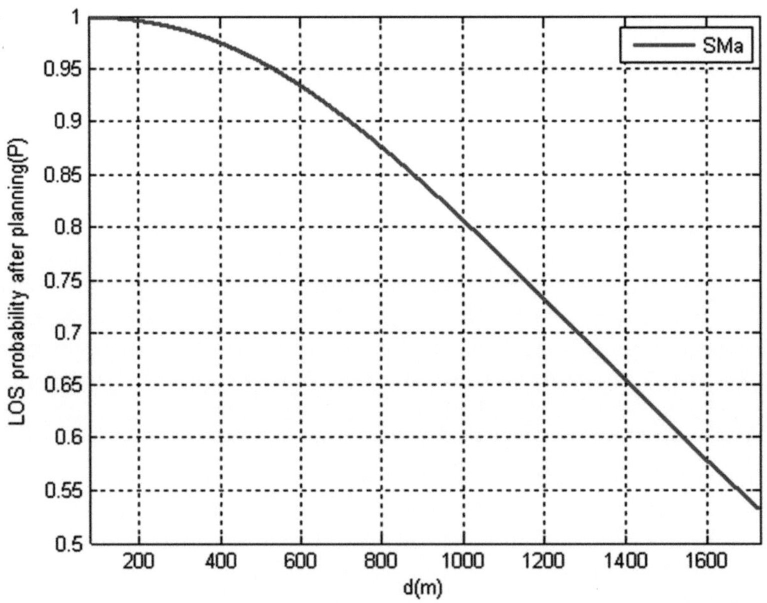

Fig. 2.21 Post site planning LOS probability vs. distance of donor eNB–RN, Suburban

Ideally, different bonuses would be applied for each individual setting. However, that would complicate the channel modeling and result in very cubersome set of parameters. Therefore, a single bonus value is perferred. From Tables 2.2 and 2.3, it is seen that using a single value could lead to approximate $1 \sim 2$ dB error when calculating the pathloss.

Alternatively, site planning can be performed in each simulation run, if a single value of bonus is considered not accurate enough to capture the actual SINR gain in the backhaul link. In this case, the following procedure can be carried out in relay system-level simulations.

The site planning optimization is a process of finding an optimal location among N candidate relay sites around the virtual relay which offers bonus to the performance.

- The relay geography locations are initialized in a system-level simulation by random dropping.
- N = 5 candidate relay sites are considered within a searching area of 50 m radius around the virtual relay.
- The best relay site is selected based on SINR criteria on the backhaul link.

2.3.2 Improvement of LOS Probability in Donor eNB–RN Connection

As the pathloss of LOS environment is significantly smaller than that of NLOS environment, a direct consequence of relay site planning is the increased chance of LOS propagation in the backhaul link.

The probability of finding a site with LOS propagation to its donor eNB depends on two factors: the LOS probability of each candidate site and the correlation between these sites. The first factor and the corresponding formulae are already discussed in previous sections. The rest is to model LOS correlation between these candidate sites. Assuming N candidate relay sites and using b_i to represent a Boolean variable indicating whether the ith candidate site is of LOS ($b_i = 1$) propagation or of NLOS ($b_i = 0$) propagation. The Boolean variable b_i can be generated from spatially-correlated Gaussian random variables g_i [7], the same way the shadow fading is generated where de-correlation distance of $d_{\text{cor\_LOS}}$ is used. The de-correlation distance captures the correlation between the two Gaussian variables at two sites. The correlation is modeled as an exponential decaying function of distance Δx. It is known that the propagation mechanisms of shadow fading and LOS are fundamentally similar, both heavily influenced by the buildings, scatterers, and/or obstacles. Therefore, we can assume that for the same eNB to RN connection, the correlation model and the de-correlation distance are similar for shadow fading and LOS probability, i.e., $d_{\text{cor\_LOS}} = d_{\text{cor\_SF}} = 50$ m. Denoting g_{i} as the Gaussian random variable for the ith candidate site, b_i can be calculated as follows

Table 2.4 Large scale fading parameters for outdoor relay performance evaluation

Distance-dependent path loss	Macro to relay: $PL_{LOS}(R) = 100.7 + 23.5log10(R)$ $PL_{NLOS}(R) = 125.2 + 36.3log10(R)$ For 2 GHz, *R* in km. LOS probability: Case 1: $Prob(R) = min(0.018/R,1)*(1-exp(-R/0.072)) + exp(-R/0.072)$ Case 3 (*Suburban*): $Prob(R) = exp(-(R-0.01)/0.23)$ Case 3 (*Rural/Suburban*): $Prob(R) = exp(-(R-0.01)/1.15)$ Bonus for donor macro (from each of its sectors) to relay for optimized deployment by site planning optimization methodology. Higher probability of LOS shall be reflected in consideration of the height of RN antenna and site planning optimization. *Bonus* = 5 dB, for donor macro (from each of its sectors) to relay for NLOS, 0 dB for LOS LOS probability correction as $1-(1-Prob(R))$^N, where $N = 3$ Relay to UE: $PL_{LOS}(R) = 103.8 + 20.9log10(R)$ $PL_{NLOS}(R) = 145.4 + 37.5log10(R)$ For 2 GHz, *R* in km Case 1: $Prob(R) = 0.5-min(0.5,5exp(-0.156/R)) + min(0.5, 5exp(-R/0.03))$ Case 3: $Prob(R) = 0.5-min(0.5,3exp(-0.3/R)) + min(0.5, 3exp(-R/0.095))$
Shadowing standard deviation	Macro to relay Relay with outdoor donor antenna: 6 dB Relay to UE: Relay with outdoor coverage antenna: 10 dB

$$b = \begin{cases} 1, & \text{if } g_i < \sqrt{2} \cdot \text{erfinv}(2p - 1) \\ 0, & \text{otherwise} \end{cases}$$

where p is the LOS probability for the candidate site and erfinv(·) is the inverse error function. Note b_i are correlated in space, since g_i are correlated in space. The site selection procedure from N candidate sites can be mathematically represented as max $(b_1, b_2, \ldots, b_N) = 1$ or equivalent. So after RN site planning the LOS probability becomes

$$Prob\left(\min(g_1, g_2, \ldots, g_N) < \sqrt{2} \cdot \text{erfinv}(2p - 1)\right)$$

Considering five candidate sites within a circular area of radius 50 and 30 m and these sites are randomly placed in the searching area, the distribution of $\min(b_1, b_2, \ldots, b_5)$ is obtained from the simulation. Then, the post-planning LOS probability is calculated as a function of the pre-planning LOS probability. In [7] it is shown that $Prob = 1-(1-p)^{3.1}$ matches well the simulated post-planning LOS probability for 50 m search radius. To simplify the formula, the exponential "3.1" was round-off to the nearest integer 3 in the final agreement for the modeling of post-site-planning LOS probability. Figures 2.20 and 2.21 show LOS probabilities as a function of the distance between donor eNB and RN, after site planning, for Dense Urban and Suburban environment, respectively.

2.4 Large Scale Fading Parameters

Large scale fading parameters for eNB–RN and RN–UE connections are summarized in Table 2.4. The relay here is supposed to be outdoor relay. The effect of relay site optimization is also included in Table 2.4.

Besides the pathloss, shadow fading is another important aspect for channel modeling. In WINNER channel model, the shadow fading of LOS and NLOS are different. And for indoor users in UMi, the more shadowing is applied. For relay channels, due to the lack of time for detail analysis of the measurement data, there is no differentiation between LOS and NLOS in terms of shadow fading. For macro-UE connection, the shadow fading standard deviation is often set to be 8–8.9 dB. Considering the higher antennas at RN compared to UE, smaller standard deviation is expected for backhaul link, which is modeled as 6 dB for outdoor relay [8]. On the other hand, the lower antenna height of RN compared to macro eNB leads to higher standard deviation, i.e., 10 dB for RN to UE connection.

2.5 Small Scale Fading

Most effort of channel modeling in Release 10 relay study item was focused on large-scale fading modeling which fundamentally determines the performance expectation of relay systems.

Nevertheless, fast fading is also important in the sense that:

- Delay spread, or frequency domain characteristics can only be captured by the fast fading. Frequency selective scheduling is a key feature of OFDM system which would significantly improve the system performance. Without modeling the fast fading, simulation results would only reflect the flat fading performance which effectively disables the frequency selective scheduling.
- Small scale time domain statistics are crucial to those transmission schemes that rely on the feedback of channel state information (CSI). The static channel assumption in large scale fading only simulations would exaggerate the actual performance.
- Multi-antenna technologies rely heavily on the assumptions of spatial channel characteristics, such as angle of arrival (AoA), angle of departure (AoD), angle spread, cross-polarization discrimination, which can only be obtained by fast fading modeling.

Table 2.5 Delay spread, angle spread parameters for relay-UE and macro-relay connections, compared with ITU UMa and ITU UMi

Scenarios/links	Value					
	Average delay spread (μs)		Average spread of AoD (degrees)		Average spread of AoA (degrees)	
	LOS	NLOS	LOS	NLOS	LOS	NLOS
Relay-UE	0.089	0.148	18	30	25	36
Macro-relay	0.079	0.174	23	29	35	40
ITU UMa	0.093	0.363	14	26	65	74
ITU UMi	0.065	0.129	16	26	56	69

Note that, for simplicity of modeling, absolute value of cross-correlation smaller than 0.3 is set to 0

- Link level simulations require elaborate modeling of small scale fading

Several agreed models for fast fading, such as SCM, SCM-E, or ITU UMa/UMi, could be used in the absence of well accepted models for relay backhaul and access links. But those models may not accurately represent the fast fading characteristics of eNB–RN and RN–UE links, as seen shortly after.

There were some proposals on fast fading modeling for relay-UE and eNB-relay connections as [9]. The proposed parameters were based on the data from the similar measurement campaign as for pathloss modeling discussed in Sect. 2.2. The methodology follows the same procedures of ITU fast fading modeling, i.e., using the same formulae, for example,

- The delay spread (DS) distribution is still modeled as exponential decaying function with certain mean value to represent the overall channel frequency selectivity. The standard deviation is to capture the randomness of delay profile of each UE in various locations.
- The azimuth angle spread distribution is modeled as wrapped Gaussian, for both angle of departure (AoD) and angle of arrival (AoA). In relay-UE link, the angle spreads of AoD and AoA reflect the spatial richness from the aspects of relay access link antennas, and UE antennas, respectively. In macro-relay link, the angle spreads of AoD and AoA capture the characteristics scatterers from the aspects of macro antennas and relay backhaul link antennas, respectively.
- The composite parameters for delay spread, azimuth angle spread and shadow fading (SF) are correlated. Six cross-correlation coefficients are used: angle spread of departure (ASD) to delay spread, angle spread of arrival (ASA) to delay spread, ASD to shadowing, ASA to shadowing, ASD to ADA, delay spread to shadowing.

The above fast fading parameters for relay-UE and macro-relay links are proposed in [10] to better match the channel measurement data. Tables 2.5 and 2.6 highlight some key parameters, compared with those of ITU UMa and ITU UMi.

In Table 2.5, the average delay spread for LOS does not differ much between relay channels and ITU macro and micro channels. However, For NLOS, the delay spread of ITU UMa is significantly longer that other channels. It is also observed

Table 2.6 Cross correlations between delay spread, angle spread parameters for relay-UE and macro-relay connections, compared with ITU UMa and ITU UMi

Scenarios/ links	Cross correlation											
	ASD to DS		ASA to DS		ASA to SF		ASD to SF		DS to SF		ASA to ASD	
	LOS	NLOS	LOS	NLOS	LOS	NLOS	LOS	NLOS	LOS	NLOS	LOS	NLOS
RN–UE	0.3	0.3	0.5	0	0	0	0	0	−0.8	−0.8	0	0.5
eNB-RN	0	0.6	0.3	0	−0.3	0	0	0	−0.3	−0.3	0	0
ITU UMa	0.4	0.4	0.8	0.6	−0.5	0	−0.5	−0.6	−0.4	−0.4	0	0.4
ITU UMi	0.5	0	0.8	0.4	−0.4	-0.4	−0.5	0	−0.4	−0.7	0.4	0

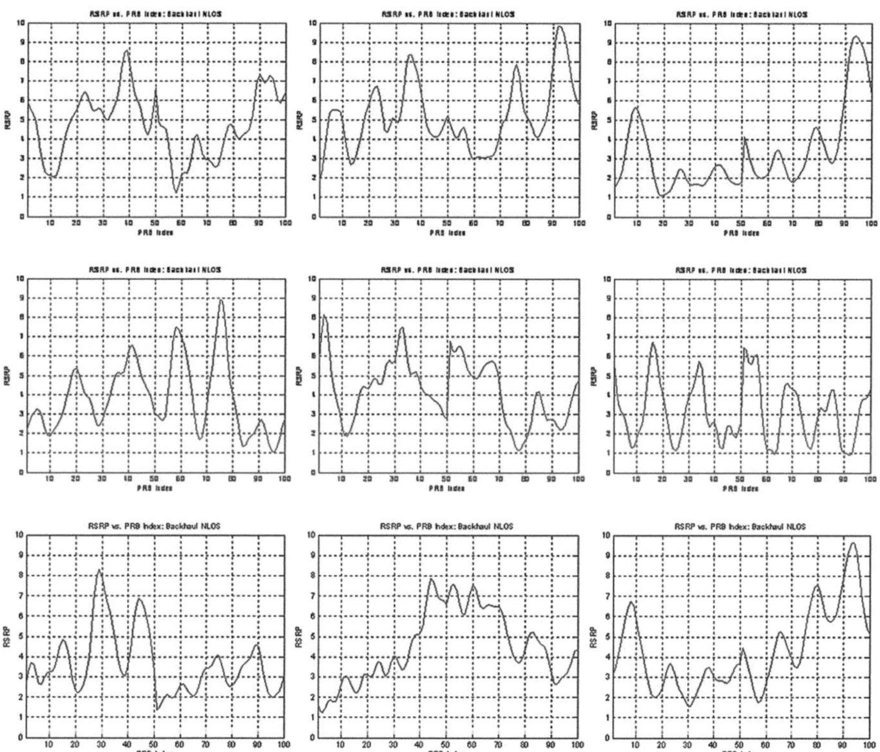

Fig. 2.22 RSRP vs. PRB index of backhaul channels in NLOS scenario

that due to the higher elevation of relay antennas compared to UE antennas, angle spread of AoA of backhaul link is noticeably narrower than that of ITU UMa and ITU UMi, which means from relay backhaul antenna point of view, less scatterers are seen. It is interesting to see that angle spread of AoA of access link is narrower than that of ITU UMa and ITU UMi, which may be explained by the wave-guide

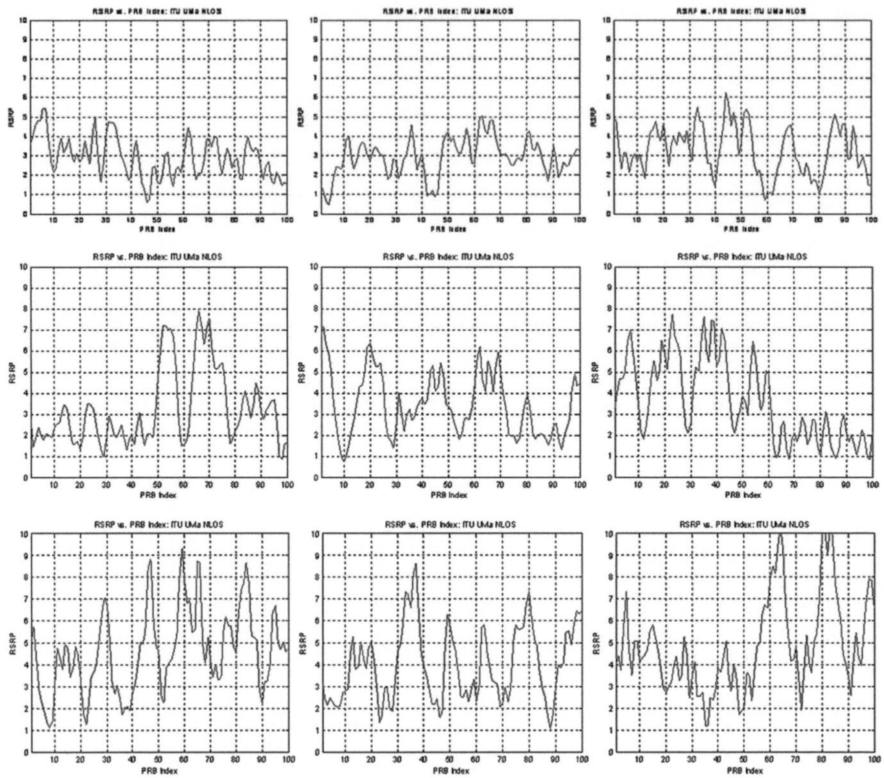

Fig. 2.23 RSRP vs. PRB index of ITU UMa channels in NLOS scenario

effect, caused by the lower RN antenna height than macro or micro eNB. Also observed is the slightly broader angle spread of AoD in eNB–RN link of LOS propagation. Possible reason is that eNB–RN link would receive signals from far-away scatters than RN–UE and eNB–UE links.

Table 2.6 shows that in all the four scenarios/links, angle spread and delay spread are moderate-high positive correlated, and delay spread and shadow fading are moderate-high negative correlated. In ITU UMa and ITU UMi, the angle spread and shadow fading are moderate-high negative correlated. However in RN–UE and eNB–RN links, the correlation between the angle spread and shadow fading is small.

The parameters in Table 2.5 may look obscure to people who are not familiar with spatial channel modeling or ITU fast fading channel models. To get a feeling of what the channel looks like with Table 2.5, we plot a few frequency domain response of the backhaul channel using those parameters.

We randomly generate 9 fast fading realizations and plot linear-scale reference signal received power (RSRP) as a function of physical resource block (PRB) index. The operating bandwidth is 20 MHz, and there are 100 PRBs in

Table 2.7 Other key parameters for outdoor relay system simulations

Parameter	Assumption/value
Cell layout	Hexagonal grid, 19 macro eNB cell sites, 3 cells per site, wrapped around
Inter-site distance (macro)	500 m (Case 1), 1,732 m (Case 3)
Penetration loss	Macro to UE: 20 dB, macro to RN: 0 dB, RN to UE: 20 dB
Antenna pattern (azimuth)	Relay-UE link (Case 1): 5dBi antenna gain, omni $A(\theta) = 0$ dB 2 transmit, 2 receive antenna configuration Relay-UE link (Case 3): 5dBi antenna gain, omni $A(\theta) = 0$ dB or directional pointing away from the donor cell $A(\theta) = -\min\left[12\left(\frac{\theta}{\theta_{3db}}\right)^2, A_m\right]$ $\theta_{3dB} = 70$ degrees, $A_m = 20$ dB. 2 transmit, 2 receive antenna configuration Macro-Relay link (Case 1 and Case 3) 7dBi, directional $A(\theta) = -\min\left[12\left(\frac{\theta}{\theta_{3db}}\right)^2, A_m\right]$ $\theta_{3dB} = 70$ degrees, $A_m = 20$ dB. 2 transmit, 2 receive antenna configuration, or 4 transmit, 4 receive antenna configuration Use of antenna downtilt and vertical antenna FFS
Total transmit power of RN	Case 1: 30 dBm @ 10 MHz bandwidth Case 3: 30 or 37 dBm @ 10 MHz bandwidth
Mini dist. between UE and outdoor RN	10 m
Mini dist between RN and marco	35 m
Num of UEs per macro cell	25

the plots. RSRPs in each figure are the averaged values over resource elements of common reference signal (CRS) in each PRB. Figure 2.22 corresponds to backhaul NLOS scenario. Figure 2.23 is for ITU UMa NLOS scenario. It is seen that in NLOS scenario the backhaul channel frequency response is smoother than ITU UMa channels, indicating that the backhaul channel is less frequency selective than ITU UMa channel. This is reasonable according to Table 2.5 where the average delay spread of macro-RN channel is 0.174 μs, compared to 0.363 μs for ITU UMa.Due to the limited time of the simulation verification, no new fast fading models were agreed for relay study. Nevertheless, the parameters proposed in [10] would be a useful reference for future study of relay, and even for pico cell (access link).

2.6 Other Settings

Relay system simulation reuse many parameters for homogeneous network simulations. But there are some exceptions such as the transmit power of RN, minimum distance between UE and RN, between RN and macro, etc. Some key parameters are listed in Table 2.7.

References

1. 3GPP R1-082975: Application scenarios for LTE-Advanced relay, RAN1 #54, China Mobile, Vodafone, Huawei, Aug 2008,
2. 3GPP R1-094303: Support of indoor relays in LTE-Advanced, RAN1 #58bis, Orange, Oct 2009
3. 3GPP TS 306.104: Base station (BS) radio transmission and reception
4. 3GPP R1-091566: Relay to UE channel model for LTE-Advanced, RAN1 #56bis, CMCC, Mar 2009
5. 3GPP R1-092820: Proposal for macro-relay pathloss model based on measurement results, RAN1 #57bis, CMCC, June 2009
6. 3GPP R1-093271: Consideration of site planning on relay evaluation methodology, RAN1 #58, CMCC, Aug 2009
7. 3GPP R1-093486: Impact of relay site planning on LOS probability of the backhaul link, RAN1 #58, Ericsson, ST-Ericsson, Aug 2009
8. 3GPP TR 36.814: Evolved Universal Terrestrial Radio Access (E-UTRA): Further advancements for E-UTRA physical layer aspects.
9. 3GPP R1-100559: Further consideration on relay channel modeling, RAN1 #59bis, CMCC, Jan 2010
10. ITU-R M.2135: Guidelines for evaluation of radio interface technologies for IMT-Advanced

Chapter 3
LTE-A Relay Study and Related Technologies

3GPP study item phase is a relatively open discussion period when various potential techniques can be proposed. In LTE-A relay study item, technical proposals encompass the resource partitions between hops, relay categorizations, inband versus outband operations. The effort was later narrowed down to two major relay types: Type 1 relay and Type 2 relay, during which companies conducted extensive simulation evaluations and proposed high-level design guidelines.

Relay study activities were not limited to the time frame of LTE-A study item which began in January and ended in December 2010. Some research topics, for example carrier aggregation for the backhaul link, continued into the work item of LTE-A, although not gaining much attentions. Outside 3GPP, there has been quite a lot of research on the carrier aggregation operation and multi-hops (>2) for relay [1]. These potential techniques, though not specified in LTE Release 10, are valuable references for future relay work.

There are a few inter-connections between relay and other LTE-A technologies, where the study and specification of one technology would affect the other. To facilitate the understanding of the contents in next few chapters, we also briefly describe the related topics in LTE-A: carrier aggregation, downlink reference signals, enhanced ICIC and CoMP.

3.1 Relay Categorization Based on Protocol Architecture

The most straightforward way to categorize a relay node is by checking the protocol architecture. From that prospective, three types can be defined: Layer 1 (L1), Layer 2 (L2) and Layer 3 (L3) [2]. As the understanding went deeper regarding how relaying would actually operate, pure protocol layer based categorization lost it significance in RAN1, and were replaced by more relevant relay types such as Type 1 relay and Type 2 relay. Nevertheless, the discussion of layer protocols for relay is helpful in understanding the basic operations of a relay node and higher layer aspect of relays.

Y. Yuan, *LTE-Advanced Relay Technology and Standardization*,
Signals and Communication Technology, DOI: 10.1007/978-3-642-29676-5_3,

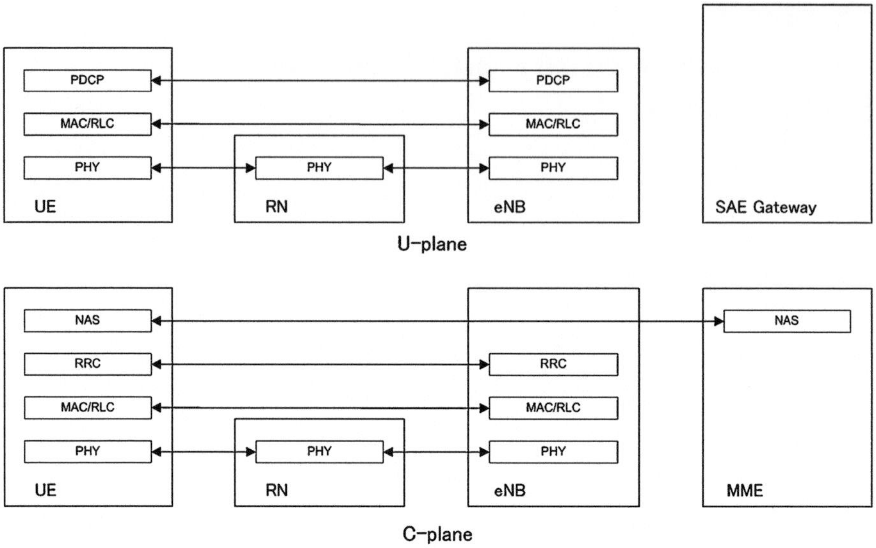

Fig. 3.1 Protocol stack for L1 relay node

3.1.1 L1 Relay

L1 relay features "amplify-and-forward". It simply amplifies the signal (presumably mixed with the interference/noise) from eNB/UE and forward it to the UE/eNB. A common form of L1 relay is the so called "repeater" that has no baseband processing at all, i.e., the signal is amplified and forwarded across the entire bandwidth. A more sophisticated repeater can have frequency-selection capability, i.e., it would only amplify-and-forward the signal in the target frequency resources, thus causing less interference to the system. The frequency-selective L1 relaying can be achieved purely in RF domain or requires some baseband processing, depending the actual implementation. Overall, L1 relay has the least processing delay, which may support full-duplex transmission of the RN. The protocol stack of L1 relay in U-plane and C-plane is shown in Fig. 3.1.

3.1.2 L2 Relay

L2 relay has MAC layer, and possibly RLC layer also. It has its own scheduler and can support multiplex, de-multiplex of MAC SDU and priority handling between RN and UE. The radio resource allocation between UE and RN is carried out in coordination with eNB and the possibly with other RNs by taking into account the inter-cell interference and load condition. In addition, the outer ARQ and RLC PDU segmentation/concatenation might reside in RN. The protocol stack of L2 relay is illustrated in Fig. 3.2.

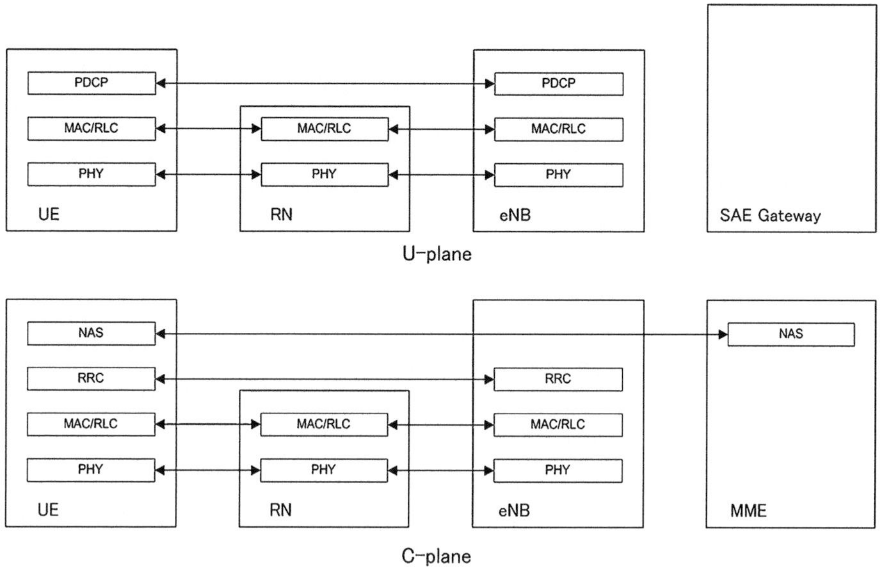

Fig. 3.2 Protocol stack for L2 relay node

3.1.3 L3 Relay

L3 relay has partial or full functions of RRC in eNB as seen in Fig. 3.3. As RRCs are distributed in macro and relay nodes, there are more handover scenarios than the current E-UTRAN, e.g., eNB–eNB, eNB-RN (in the same eNB), eNB-RN (in the different eNB), RN–RN (in the same eNB), RN–RN (in the different eNB). The L3 measurement may be reused for handover decision in RN.

Table 3.1 shows the functional allocation to different types of RNs, compared to eNB. For L1 RN, radio resource scheduling is performed in a centralized fashion by the eNB. L2 RN and L3 RNs have their own schedulers to assign the resources for UEs within the relay coverage.

The pros and cons for the above three RN types are compared below.

- L1 relay

  Pros:
  L1 relay can improve cell coverage, possibility of SFN combining without coordination. It has the smallest delay compared to L2 relay and L3 relay.

  Cons:
  Noise is also amplified in the physical link. There is potential loop interference between transmit and receive antennas if they are sufficiently isolated. L1 relay cannot improve the spectral efficiency.

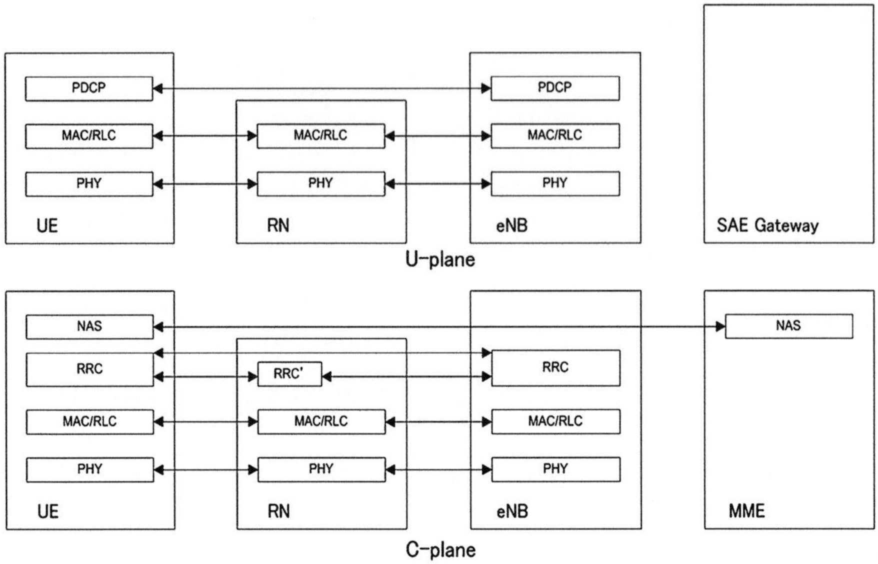

Fig. 3.3 Protocol stack for L3 relay node

Table 3.1 Functional allocation in relay node

Function	L1 RN	L2 RN	L3 RN	eNB
RF function	X	X	X	X
Coder/decoder and CRC	–	X	X	X
HARQ	–	X	X	X
Multiplex and de-multiplex of MAC SDU	–	X	X	X
Priority (QoS) handling	–	X	X	X
Scheduling	–	X	X	X
Outer ARQ	–	X	X	X
(Re)-Segmentation and concatenation	–	X	X	X
System information broadcast	–	–	X	X
RRC connection set-up and maintenance	–	–	X	X
Mobility function	–	–	X	X
QoS management	–	–	X	X
UE measurement and control reporting	–	–	X	X
Header compression (ROHC)	–	–	–	X
Reordering of lower layer SDUs	–	–	–	X
In-sequence delivery of upper layer PDUs	–	–	–	X
Duplicate detection of lower layer SDUs	–	–	–	X
Ciphering	–	–	–	X
Radio bearers set-up and maintenance	–	–	–	X
MBMS services control	–	–	–	X
Paging	–	–	–	X
NAS signaling handling	–	–	–	X

- L2 relay

  Pros:
  L2 relay can improve cell coverage. Due to the decode-and-forward nature, L2 relay can reduce the error rate on eNB-RN and RN-UE links via HARQ based error correction. L2 relay can have dynamic priority handling and scheduling to adapt to the channel condition in each RN. Fast adaptive modulation and coding can be supported for RN-UE link. L2 relay allows flexible RLC segmentation and concatenation when the transport block sizes of eNB-RN and RN-UE links are different.

  Cons:
  Residual error after HARQ causes radio resource waste. It is difficult to perform SFN combining without coordination between L2 relay nodes, or between L2 relay node and donor eNB.

- L3 relay

  Pros:
  In addition to the benefits of L2 relay, L3 relay can reduce the latency on RRC connection set up and handover procedure.

  Cons:
  Besides the disadvantages of L2 relay, increased latency is expected from the source node to the destination node when L3 is used, compared to L1 and L2 relays.

3.2 Operating Band

Band allocation between relay hops is a fundamental aspect of relay operation. Simply speaking, the operation can be either inband, i.e., eNB-RN link and RN-UE link share the same band, or outband, i.e., eNB-RN link and RN-UE link are in the different band.

Outband relay has the benefit of full duplex operation, with no change to Release 8 UE/eNB behavior. In another word, standard effort for outband relay is minimal from physical layer specification point of view. The drawback of outband relay is that it requires additional spectrum for the backhaul link, and the frequency bands of access link and backhaul links have to be sufficiently separated. Adjacent carriers may not provide adequate frequency spacing to block the loop coupling. The hardware cost of outband relay is expected to be high due to the additional duplexer to support different bands.

It is noted that full duplex can still be achieved in decode-and-forward inband relay if the RN's transmit antenna and receive antenna are sufficiently isolated. In another word, the backhaul link and the access link can be spatially isolated. Similar to the case of outband relay, such spatial-separation based relay case requires minimum specification effort in RAN1.

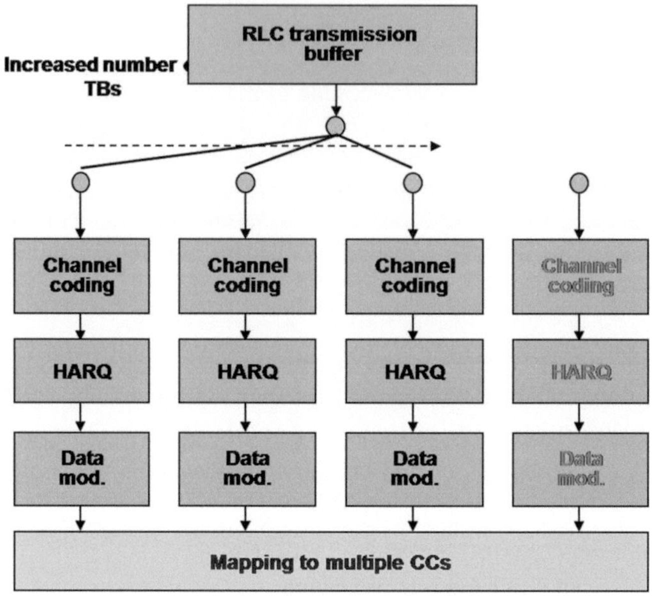

Fig. 3.4 MAC layer mapping from transport blocks to component carriers

The study on carrier aggregation has some implications on outband relaying, a decode-and-forward relay with carrier aggregation capability can achieve "outband" operation by simultaneously scheduling different component carriers in the backhaul link and the access link. Some investigations show the performance benefits of carrier aggregation in the backhaul [1] which is to be further discussed shortly after. In 3GPP RAN1, carrier aggregation aspect for relay did not gain much interest, possibly due to: (1) the cost consideration for relay node to support CA in the backhaul; (2) significant specification work to support CA features in relay study and work items.

Band operation also has some relation to the discussion of resource partitions between for example backhaul link (donor eNB-RN) and access link (RN- UE).

3.2.1 Brief Description of LTE-A Carrier Aggregation

Carrier aggregation (CA) is supported in both FDD and TDD for Release 10. In FDD downlink, both intra-band and inter-band aggregations are allowed. However, intra-band aggregation is prioritized in FDD uplink. For TDD, intra-band aggregation is prioritized in both downlink and uplink. In Release 10, the uplink carrier aggregation specification is optimized for 40 MHz, i.e., two component carrier aggregation.

As illustrated in Fig. 3.4, transport blocks from RLC layer go through multiple independent HARQs, each uniquely mapped to one component carrier.

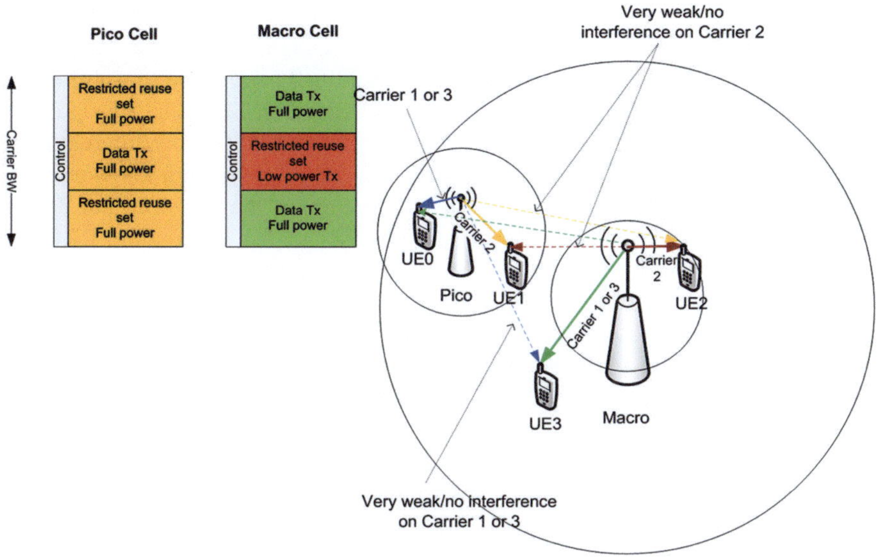

Fig. 3.5 Cross-carrier scheduling in heterogeneous deployment

As a new feature in Release 10, PDCCH on a component carrier can assign PDSCH or PUSCH resources in one of multiple component carriers using the carrier indicator field (CIF). The configuration for the presence of CIF is semi-static and UE specific. CIF contains 3-bit information and is jointly coded with Release 8 PDCCH structure. CIF location is fixed irrespective of DCI format size. Cross-carrier assignments can be configured when the DCI formats have the same or different sizes. Explicit CIF is used at least for the case of same DCI format size. CIF mapping to component carriers is UE specific and configured by RRC.

One key motivation of cross carrier scheduling is to facilitate inter-cell interference coordination (ICIC) in heterogeneous deployment as illustrated in Fig. 3.5. The coverage of pico node is significantly smaller than that of macro cell due to the relatively low transmit power, low antenna gain and lower height of antenna deployment. Therefore, all three component carriers can be turned on for the pico cell without causing too much interference to UEs in macro cell coverage. For example, UE3 can still reliably receive the signals from macro cell in Carrier 1 or Carrier 3, as long as the macro cell can apply full transmit power in those two carriers. Although the pico may transmit data on all three carriers, the strong potential interference from macro cell would hamper the spectral efficiency on carriers that are not protected, i.e., Carrier 1 and Carrier 3. In another word, only Carrier 2 is fully in the hand of the pico for reliable high speed transmission, i.e., macro configures Carrier 2 as restricted carrier that with no data transmission or operating at very low power. Since the interference happens most at PDSCH region, macro cell can still transmit PDCCH in the first few OFDM symbols of a subframe in Carrier 2, but leaving the subsequent PDSCH region empty.

Therefore, it would be quite useful if the PDCCH in Carrier 2 can cross-schedule the PDSCH in Carrier 1 and/or Carrier 3, which so called cross-carrier scheduling.

Normally, the PDCCH region may vary dynamically subframe by subframe to fit the ever changing demand for control channel capacity. This can cause problem for cross-carrier scheduling, as the starting symbol of PDSCH on different component carriers can be different. In another word, the UE would not know where to start decoding PDSCH in the cross-scheduled carrier if the UE does not decode PCFICH in that carrier. Therefore, in case of cross-carrier scheduling, RRC signaling is used to inform the CFI to the UE for the carriers on which PDSCH is assigned.

Considering the heavier bit load of uplink control channels associated with carrier aggregation, in particular with TDD deployment, new PUCCH format is specified for ACK larger than 4 bits. The new format supports up to 21 ACK/NACK feedback bits in TDD.

DFT-precoded OFDM is the transmission scheme for PUSCH both in the absence and with the presence of spatial multiplexing. In the case of multiple component carriers, there is one DFT per component carrier. The DCI format supports both frequency-contiguous and frequency non-contiguous resource allocation in each component carrier.

LTE-A supports component carrier specific UL power control for both contiguous and non-contiguous carrier aggregation for closed-loop case, and for open loop at least for the cases that the number of downlink component carriers is more than or equal to that of uplink component carriers.

3.2.2 Relay with Carrier Aggregation

Several scenarios can be considered for carrier aggregation for relay as illustrated in Fig. 3.6 [3].

In relay CA Scenario 1, the component carriers configured on the backhaul link are all different from those on the access link. In another word, it is a pure outband relay. Therefore, minimal effort is expected in physical layer specification. However, performance requirement still needs to be specified by RAN4 working group.

In relay CA Scenario 2, one component carrier, e.g., CC#1, is configured on both the backhaul and access links. Clearly, backhaul/access link subframes must be configured for CC#1 for inband operation. It is reasonable to assume that the signaling and procedures defined for single carrier inband relay operation can be reused for this purpose, as long as CC#1 is configured as the primary component carrier (PCC). It is preferred not to configure CC#1 as a secondary component carrier (SCC) as such configuration would potentially impact the specifications.

In relay CA Scenario 3, the component carriers configured on the backhaul link are exactly the same as those on the access link. Hence, the normal signaling and procedures defined for carrier can be reused, without further standardization tuned

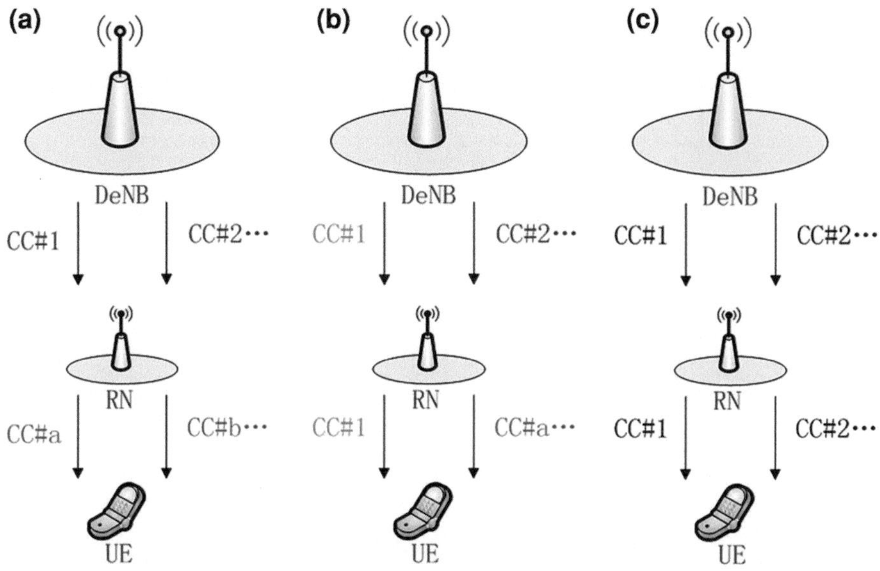

Fig. 3.6 Three scenarios of relay with carrier aggregation capability. **a** Relay CA scenario 1, **b** relay CA scenario 2, **c** relay CA scenario 3

for relay CA operation specifically. But similar to relay CA Scenario 1, performance requirement has to be specified by RAN4.

One important aspect of performance requirement associated with carrier aggregation is the loop interference between the backhaul link and the access link. Some performance study was carried out to evaluate that impact [1]. Single-band carrier aggregation is of concern where component carriers are allocated within a block of band and the cross-carrier leakage is often not to be neglected. The following two cases were considered:

- Case 1 (inter site distance = 500 m)

  Received backhaul signal power: −44 dBm,
  Backhaul link signal quality SINR = 16 dB

- Case 3 (inter site distance = 1,732 m)

  Received backhaul signal power: −64 dBm,
  Backhaul link signal quality SINR = 16 dB

The transmit power density of relay node is 13 dBm per PRB. The adjacent carrier interference rejection (ACIR) at relay node receiver is 42.5 dB.

Figure 3.7 shows the degradation of received SINR in the backhaul as a function of antenna isolation levels between the backhaul link antenna and the access link antenna. Obviously, the weaker received power in Case 3 (20 dB lower than that in Case 1) leaves it more vulnerable to the inter-carrier loop interference. To keep the SINR degradation within 0.5 dB for the relay backhaul link, 40 dB of antenna isolation is required in Case 1 and 60 dB for Case 3.

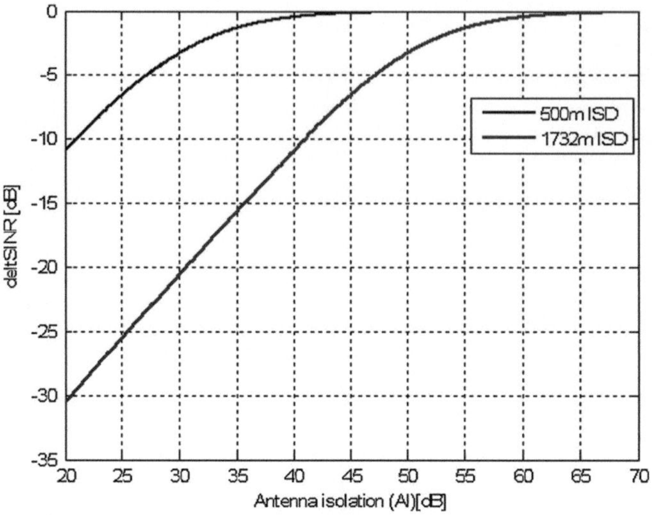

Fig. 3.7 Outband RN backhaul link signal quality degradation due to inter-carrier loop interference

In addition to the link level performance analysis, system level simulations were also conducted to evaluate the impact of inter-carrier loop interference on the relay system throughput with carrier aggregation [1]. Three modes of operation are configured:

- In-band: with relay node's backhaul and access links operated on the same carriers and time multiplexed
  - With FDM: each relay node uses only one component carrier. Component carriers are distributed to relay nodes
  - With CA: each relay node is utilizing both component carriers
- Out-band: with relay backhaul link operated on one component carrier and relay access link operated on the other component carrier
  - Without FDM: all relay nodes use the same component carrier for backhaul link operation
  - With FDM: RNs use different component carriers for backhaul link operation
- Hybrid mode: with relay backhaul link operated on one component carrier with the relay access link in the in-band mode, and with a secondary backhaul link carrier operated on the other component carrier (out-band in relation to the relay access link)
  - Without FDM: all relay nodes use the same component carrier for in-band backhaul link operation
  - With FDM: relay nodes use different component carriers for in-band backhaul link operation

Table 3.2 Relay performance with carrier aggregation capability

10 RNs/sector, 1,732 m ISD			
Gain over macro-only (reference) scenario (%)			
	Mean UE thp	5 %-tile UE thp	Jain's index
In-band FDM	27	58	7
In-band CA	28	60	8
Out-band FDM(perfect AI)	40	117	21
Out-band FDM (60 dB AI)	34	96	20
Out-band non-FDM (perfect AI)	32	60	3
Out-band non-FDM (60 dB AI)	29	50	2.8
Hybrid FDM (perfect AI)	33	102	20
Hybrid FDM (60 dB AI)	30	89	19.1
Hybrid non-FDM (perfect AI)	35	58	7.3
Hybrid non-FDM (60 dB AI)	22	40	7

MBSFN overhead is counted in all inband configurations. The outband configurations experience inter-carrier loop interference, and the performance loss due the interference is modelled as a function of the antenna isolation at relay node. The hybrid configurations suffer both inband MBSFN overhead and outband loop interference effects, but having more resources for the relay backhaul link transmission.

There are 10 relay nodes in each macro sector deployed near the cell edges in two tiers. The relay locations are randomly shifted away from the planned positions to reflect the real deployment (10 m standard deviation). Two-dimensional antenna pattern is assumed. There are two component carriers considered, each with 5 MHz bandwidth

The simulated system performances [1] are compared in Table 3.2. When the loop interference protection is perfect, the outband configurations exhibit the highest performance, and the inband configurations perform the worst, due to the MBSFN overhead. The impact of MBSFN overhead on the hybrid configurations is lower than on the in-band configurations as only about 50 % of the relayed transmissions is via inband mode.

When imperfect loop interference protection, e.g., antenna isolation = 60 dB, is considered, the inband configurations show similar or better performance than the non-FDM outband. With FDM, relatively more resources are available for RN backhaul link operation, leading to some improvement in system capacity.

The discussion above shows that it is crucial to keep the inter-carrier loop interference low when carrier aggregation is used for relay operation. In fact, the CA relay discussion is quite related to time division duplex (TDD) versus subcarrier division duplex (SDD) discussion [4] during early stage of relay study item. SDD relaying is depicted in Fig. 3.8.

Note that SDD operation is different from the case of using different component carriers in the backhaul link and access link, since SDD operates at the subcarrier level, i.e., backhaul and access links still share the same component carrier. SDD relaying implies that backhaul link transmission and access link transmission can occur at the same time.

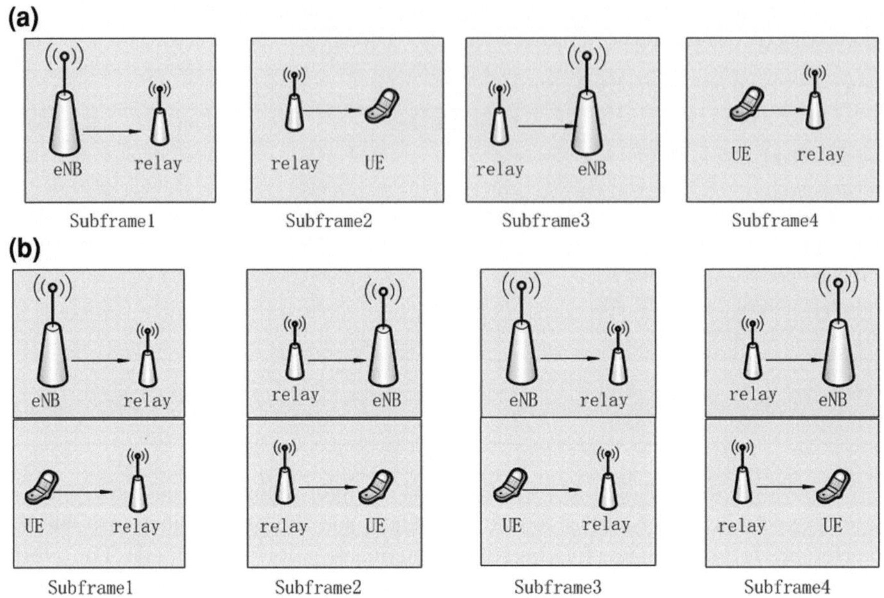

Fig. 3.8 Subframes for time division duplex versus subcarrier division duplex relay. **a** Subframe for TDD relaying, **b** subframe for SDD relaying

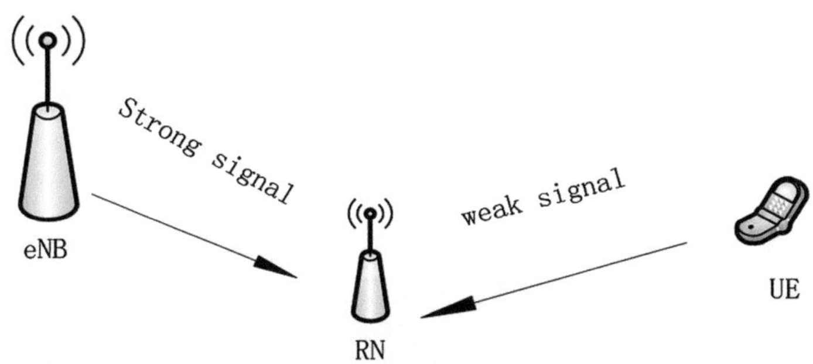

Fig. 3.9 Different received power levels experienced by relay, from eNB and UE

In an FDD system, the relay can transmit and receive simultaneously on DL and UL frequencies, where the relay can have simultaneously active backhaul and access links. In a TDD system, one frequency band is used for both DL and UL traffic, where the transmission and reception periods alternate within each eNB or UE. There are some issues with SDD relaying in the following scenarios [4].

Figure 3.9 illustrates the situation where relay experiences drastically different power levels received from the eNB and from the UE in the same OFDM symbol. Performance degradation is expected when the powers are severely imbalanced. Also, the uplink transmission of UE would de-sense the downlink reception of a neighboring UE shown in Fig. 3.10 if the two UEs are very close.

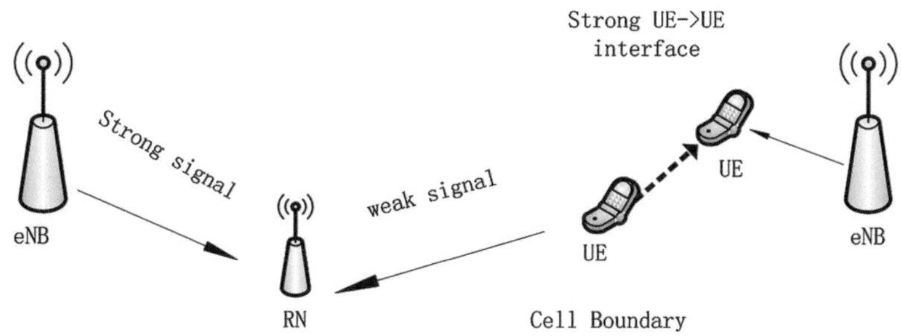

Fig. 3.10 De-sensing of downlink reception at UEs in neighboring cells

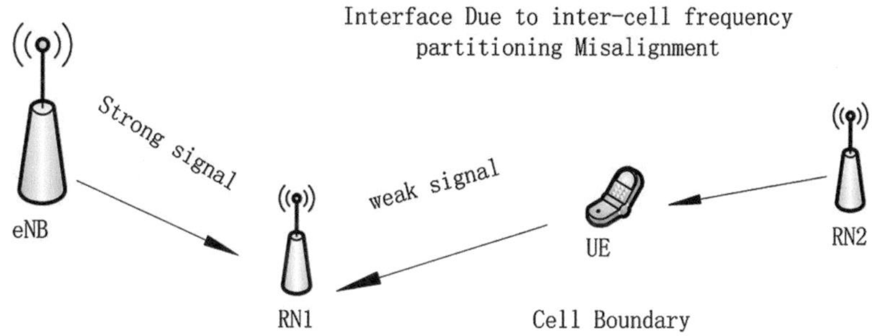

Fig. 3.11 Strong interference at UE due to misalignment of frequency partitioning

SDD requires eNB to partition the frequency resources in the backhaul and access link in a dynamic and coordinated fashion. Therefore, the misalignment between neighboring eNBs would lead to the situation that eNB-relay downlink transmission would cause significant interference to the relay-UE uplink transmission in a neighboring cell. In the same way, eNB-relay uplink transmission could interfere with relay-UE downlink transmission as shown in Fig. 3.11.

Another key problem of SDD is that ideally SDD would be applied to the entire subframe so that both PDCCH and other L1 control signalling can also be partitioned between the backhaul and access links. However, the legacy control signal channels transmitted span across the entire system bandwidth, meaning that SDD is not possible over these OFDM symbols as depicted in Fig. 3.12. When the eNB-UE downlink shares the same channel as the eNB-relay DL, the UE-relay UL would have to puncture the first few OFDM symbols, leading to further complications.

In summary, the study on CA aspects of relay backhaul and access link reveals the fact that inter-carrier interference is crucial, or a limiting factor. The study itself has certain RAN4 work nature as it involves RF requirements. Hence, further specification work in RAN4 is important from this prospective.

The difficulty of achieving good RF isolations between carriers, especially the same-band carriers, prompts using time division resource partition between backhaul link and access link for relay, or more generally using almost blank subframe to deal with the inter-cell interference (to be discussed later in this

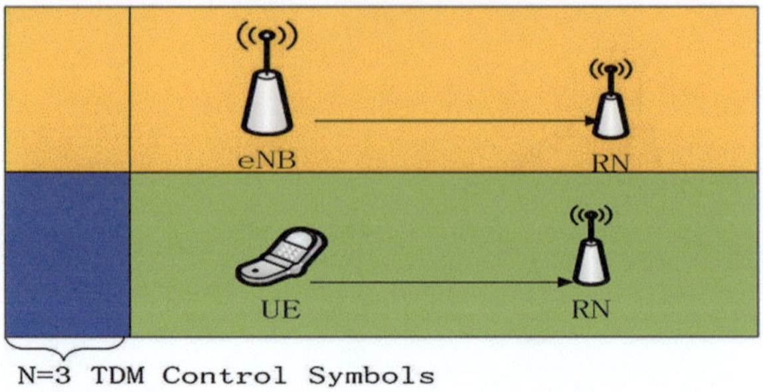

Fig. 3.12 Uplink UE-relay transmission interfered by legacy PDCCH

chapter). These TDM based approaches seem more feasible and cost effective compared to frequency based resource partitions, at least with the prevailing RF engineering in the current industry.

Frequency based resource partitions also suffer the problem of backward incompatibility with legacy PDCCH which spans over the entire system bandwidth. In this sense, legacy PDCCH seems a burden for more flexible resource allocations, which is to be further discussed in later chapters.

3.3 Number of Hops

Although most study on relay in 3GPP focuses on two-hop relays, multi-hop (>2) relays were also analyzed in ATIST 4G [1]. A tree like structure is considered, with the rationale that more relay nodes would be at lower levels than at higher levels. To minimize the bottleneck effects anticipated in the backhaul links (node to node communications), relay nodes closer to the donor eNB should be allowed to have more component carriers to provide enough backhaul capacity.

Multi-hop relay scenario makes resource management more complicated which may involve inter-node (including donor eNB-RN and RN–RN) coordination of resource assignment for interference management. The entire relay topology is nested and therefore, any resource allocation update in a relay node closer to the donor eNB would affect all relay nodes at lower levels.

For multi-hop relay systems in [1], three schemes for resource scheduling are considered: centralized, distributed, and autonomous. Centralized scheduling relies on donor eNB to coordinate all the resources in all the nodes. Relay nodes are only responsible for providing the link quality information of all involved links. In distributed scheduling, local coordination group is formed that consists of a node and its direct subordinate nodes. The autonomous resource scheduling assumes certain recognition capability of a RN that can estimate the interference level when the prospective resource configuration is used. In the simulation study in [1], the node association follows normal procedure of cell selection that is based on the reference signal received power (RSRP).

Table 3.3 lists the backhaul link (the first three hops) SINR at different percentiles when the maximum number of hops is capped to 2, 3 and 4,

Table 3.3 Backhaul link signal quality for RNs selecting multi-hop connection

1,732 m ISD, 10 RNs/sector, 49 dBm/100 MHz DeNB PSD			
SINR (dB) CDF	2-hop relaying	3-hop relaying	4-hop relaying
0.1	−16	0	6
0.2	−12	5.2	8
0.3	−9	8	11
0.4	−6	10	13.8
0.5	−4	13	14.9
0.6	−0.5	15	16
0.7	2	16	17.2
0.8	4	18	19.8
0.9	6	21.8	23
1.0	12	27	28.5

Table 3.4 Probability of multi-hop relaying

1,732 m ISD, 10 RNs/sector, 49 dBm/100 MHz DeNB PSD			
Number of backhaul link hops	Max 2-hop relaying (fraction of RNs via NH_{BH} backhaul link hops %)	Max 3-hop relaying (fraction of RNs via NH_{BH} backhaul link hops %)	Max 4-hop relaying (fraction of RNs via NH_{BH} backhaul link hops %)
1	100 %	63 %	58 %
2	–	38 %	33 %
3	–	–	9.8

respectively. Note that since relay nodes are placed near the cell edges without site optimization, the SINR of the backhaul for two-hop relay is relatively poor. It is seen that as more hops are allowed, the SINR of the backhaul is significantly increased. However, the SINR improvement diminishes as the maximum number of hops is increased from 3 to 4. The statistics of number of hops are illustrated in Table 3.4. The small percentage of 4 hops (~10 %) compared to the percentage of 3 hops (~36 % for max. 3 hops) may explain the less significant gain in backhaul SINR when the maximum number of hops is increased from 3 to 4.

Improved backhaul link SINR may not guarantee the improvement of the overall link capacity which highly depends on the resource assignment schemes and the assumption of channel state information (CSI) feedback, etc. End-to-end delay is another constraint for resource scheduling.

3.4 Type 1 Relay

While the protocol stack based categorization helped to understand the relay node functionalities, such relay category definitions were not detailed enough to describe the transmission mechanisms of a relay node, which is the main concern in physical layer study. In the meantime, amplify-and-forward type of L1 relay was no longer considered for Release 10 relay study. The focus was shifted to more detailed functionality discussions of a short list of relay node types. Among them, Type 1 relay was first identified.

3.4.1 Definition

The definition of a "type 1" relay node is:

- It control cells, each of which appears to a UE as a separate cell distinct from the donor cell
- The cells shall have its own Physical Cell ID (defined in LTE Rel-8) and the relay node shall transmit its own synchronization channels, reference symbols, …
- In the context of single-cell operation, the UE shall receive scheduling information and HARQ feedback directly from the relay node and send its control channels (SR/CQI/ACK) to the relay node
- It shall appear as a Rel-8 eNB to Rel-8 UEs (i.e., be backwards compatible)
- To LTE-A UEs, it should be possible for a type 1 relay node to appear differently than Rel-8 eNB to allow for further performance enhancement.

According the definition, type 1 relay is a L3 relay with necessary RRC functionalities to support cell handover and mobility management. Figure 3.13 is a pictorial description of type 1 relay. Such relay node is essentially a smaller-scale eNB with wireless backhaul to donor eNB. By default, a type 1 RN should have its own scheduler to allocate physical resources for UEs in its coverage zone. For backward compatibility of Release 8 UEs, type 1 RN can configure the subframes in access link to be MBSFN subframes when it receives signals from donor eNB. By doing so, Release 8 UEs served by the RN can anticipate a transmission gap in the PDSCH field.

MBSFN subframe in access link is illustrated in Fig. 3.14 where the first one or two OFDM symbols contain L1/L2 control signaling for UEs in the RN cell. Once finishing the transmission of those OFDM symbols, the RN switches to reception mode for the control signaling and data from donor eNB (DeNB). Note that RN to UE ("Uu" link) traffic is sent in normal subframes, as seen in the left part in Fig. 3.14. It should be emphasized that MBSFN subframes here are for access downlink, meaning that when an access subframe is configured as MBSFN subframe, there would be downlink control or/and data sent from DeNB to the RN. DeNB can use either normal or MBSFN subframe to serve RNs and co-schedule macro UEs simultaneously. Normal subframes may be used when a large percentage of UEs are Release 8 UEs, and MBSFN subframes may be used when most macro UEs are LTE-A UEs.

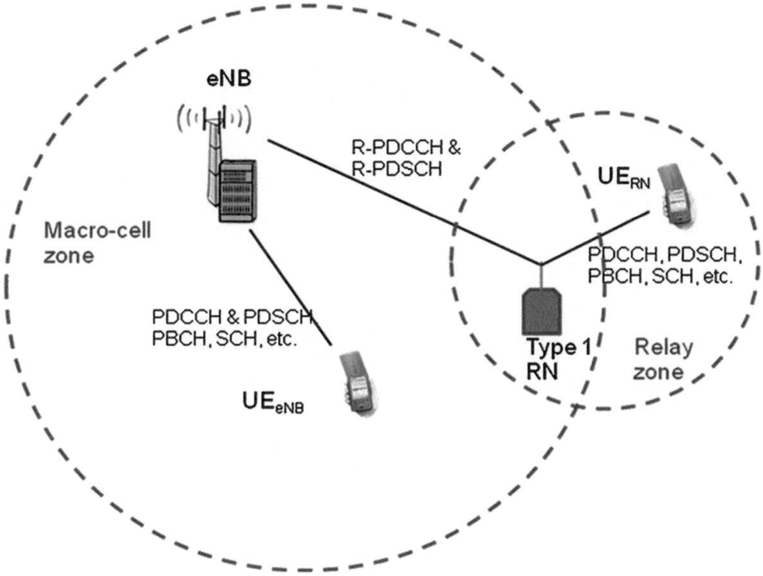

Fig. 3.13 Pictorial description of type 1 relay

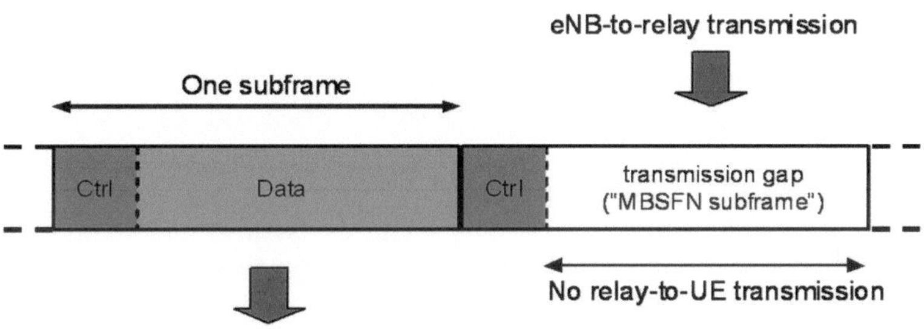

Fig. 3.14 Normal and MBSFN subframes for relay-UE link in type 1 relay

3.4.2 Technology Aspects

Since type 1 RN needs to send PDCCH to its served UEs, it cannot receive PDCCH from donor eNB. One simple solution is to reuse Release 8 PDCCH [6], but by advancing RN timing 1∼2 OFDM symbols respect to donor eNB timing counting the backhaul propagation delay, as Fig. 3.15 shows. With this timing offset, the relay node is able to receive PDCCH, without need for specifying new L1/L2 control for the backhaul. The drawback of this scheme is that the last few OFDM symbols were wasted as they cannot be received by RN. In addition, certain link adaptation scheme has to be implemented to correct the CQI report that assumes full reception of PDSCH. Another key concern is from TDD operation where generally relay node timing should be exactly synchronized with that of donor eNB. Considering the above limitations, 3GPP decided to define and optimize a new relay backhaul L1/L2 control channel called R-PDCCH.

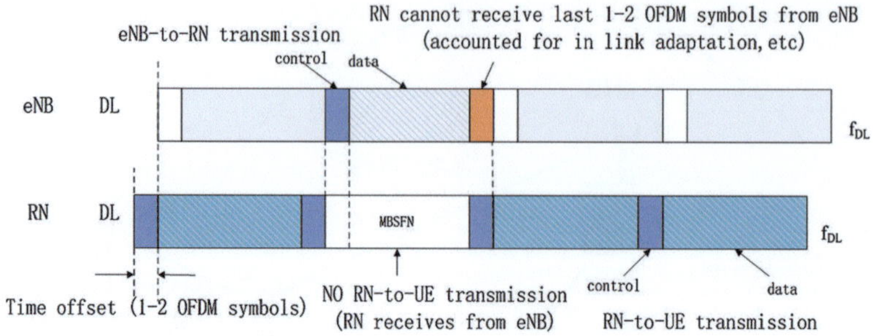

Fig. 3.15 eNB and relay node timings to reuse PDCCH for backhaul L1/L2 control

R-PDCCH would take some resources in PDSCH region. Below are some high-level agreements [5] regarding backhaul resource assignment, including for R-PDCCH. These design principles provide guidance for more detail specifications of R-PDCCH and backhaul subframe allocation as well as HARQ, to be discussed in Chap. 4.

- The set of downlink backhaul subframes is semi-statically assigned. Semi-static allocation of backhaul subframes allows relay node scheduler to optimize the dynamic resource allocation in access link subframes. Otherwise, relay node would have a moving target—the ever changing availability of subframes to optimize, which can hamper the efficient scheduling in RN cell. Such semi-static allocation may not incur significant resource waste in backhaul link even when there is no RN served in one subframe, since macro UEs can always be co-scheduled in the backhaul subframe.
- The set of uplink backhaul subframes can be semi-statically assigned, or implicitly derived from the downlink backhaul subframes, using the HARQ timing relationship. In either case, HARQ timing relationship, especially Release 8 HARQ timing should be carefully considered. HARQ collisions between access link (to follow Release 8 HARQ timing) and backhaul link should be kept minimal. In this sense, asynchronous HARQ for backhaul uplink, although allowing more flexibility of backhaul subframe allocation, is not preferred as it would cause too many HARQ collisions with the access link.
- R-PDCCH may assign downlink/uplink resources in the same and/or one or more later subframes. The motivation of assigning DL resources in later subframes is to improve the efficiency of R-PDCCH transmission, and alleviate the backhaul subframe restrictions. Due to the extensive specification work needed, there was limited interest in this approach.
- Within the PRBs semi-statically assigned for R-PDCCH transmissions, a subset of the resources is used for each R-PDCCH. The actual overall set of resources used for R-PDCCH transmission may vary dynamically between subframes. These resources may span full set of OFDM symbols available for the backhaul link or be constrained to a subset of these OFDM symbols.

- The detailed R-PDCCH transmitter processing (channel coding, interleaving, multiplexing, etc.) should reuse LTE Release 8 functionality to the extent possible, but allow removing some unnecessary procedure or bandwidth wasting procedure by considering the relay property.
- If the "search space" approach of LTE Release 8 is used for the backhaul link, use of common search space, which can be semi-statically configured (and potentially includes entire system bandwidth), is the baseline. If RN-specific search space is configured, it could be implicitly or explicitly known by RN.
- The R-PDCCH is transmitted starting from an OFDM symbol within the subframe that is late enough so that the relay can receive it. Relay frame timing will define the actual schemes which consider both Tx/Rx switching time and propagation delays of the backhaul.
- "R-PDSCH" and "R-PDCCH" can be transmitted within the same PRBs or within separated PRBs

Figure 3.16 shows an example of resource allocation that can be potentially used for R-PDCCH [7]. While the exact structure was not adopted in Release 10 relay specification, it describes the fundamental layout of R-PDCCH in frequency and time domain. Basically, R-PDCCH, shown in cyan, would occupy a few OFDM symbols, on the right side of Release 8 PDCCH region shown in green. R-PDCCH occupies partial system bandwidth, which is different from Release 8 PDCCH. The rest of frequency resources shown as the light gray areas can be used for PDSCH of macro UEs, or R-PDSCH for relay node(s). The dark gray area that follows R-PDCCH can only be used for R-PDSCH for relay node(s), since macro-UEs can neither decode R-PDCCH, nor recognize the "shortened" R-PDSCH. R-PCFICH is to dynamically signal the number of OFDM symbols for R-PDCCH, in the same way as PCFICH for PDCCH. The shadowed cyan blocks contain the signaling information for R-PDCCH frequency domain configuration.

3.4.3 Semi-Analytical Evaluations

Before showing the detailed performance results, let us first analyze the fundamental performance of type 1 relay. A simple two-hop relay setting illustrated in Fig. 3.17 can be used for the analysis.

There are two UEs considered, "UE_{in}" in the coverage area of "type 1" relay and "UE_{out}" outside of relay coverage, but served by the macro-cell. Channels are assumed flat and quasi-static between two hops. Channel coefficients (including pathloss and shadow fading) of direct link and access link are denoted as $h_{eNB\text{-}UEin}$ (or $h_{eNB\text{-}UEout}$), $h_{RN\text{-}UEin}$ (or $h_{RN\text{-}UEout}$), respectively. Transmit powers of eNB and RN are P_{eNB} and P_{RN}. To limit the scope of the study, UE receivers are not supposed to cancel the interference from other cells or other relay nodes. Other-cell interference power seen at UE_{in} and UE_{out} are denoted as $I_{o,UEin}$, $I_{o,UEout}$, respectively. We use notations $R_{eNB\text{-}RN}$, $R_{RN\text{-}UEin}$, $R_{eNB\text{-}UEin}$ and $R_{eNB\text{-}UEout}$ to

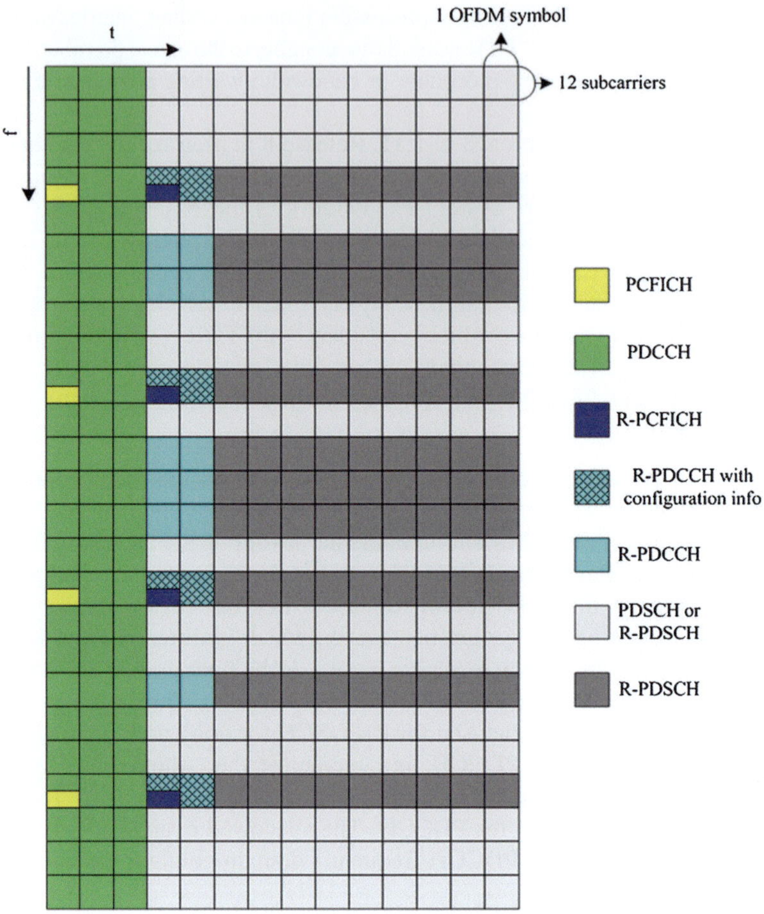

Fig. 3.16 A proposal of R-PDCCH structure potentially for type 1 relay

represent the corresponding channel capacities (assuming Gaussian signals). Here we further differentiate the representations in "type 1" relay and L2 cooperative relay, and write them down explicitly as

$$R_{\mathrm{RN-UEin,type1}} = 1 + \log_2\left[\frac{P_{\mathrm{RN}}|h_{\mathrm{RN-UEin}}|^2}{I_{\mathrm{o,UEin}} + P_{\mathrm{eNB}}|h_{\mathrm{eNB-UEin}}|^2}\right]$$

$$R_{\mathrm{RN-UEout,type1}} = 1 + \log_2\left[\frac{P_{\mathrm{eNB}}|h_{\mathrm{eNB-UEout}}|^2}{I_{\mathrm{o,UEout}} + P_{\mathrm{RN}}|h_{\mathrm{RN-UEout}}|^2}\right]$$

Note the intra-cell interference terms ($P_{\mathrm{eNB}}|h_{\mathrm{eNB\text{-}UEin}}|^2$ and $P_{\mathrm{RN}}|h_{\mathrm{RN\text{-}UEout}}|^2$) seen in "type 1" relay, cause reduced rate in access link. In fading environment, the intra-cell interference changes in time, causing wide fluctuations of CQI and SINR which generally degrade the performance of scheduling and decoding.

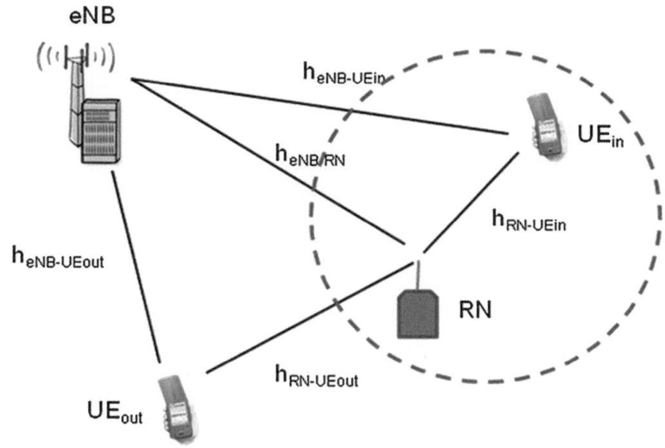

Fig. 3.17 Channels involved in a general two-hop relay setting

Assuming that resources are proportionally allocated between the two hops, the overall channel capacities can be calculated as

$$R_{\text{eff,type1}} = \frac{R_{\text{eNB-RN}} \times R_{\text{RN-UEin,type1}}}{R_{\text{eNB-RN}} + R_{\text{RN-UEin,type1}}} + R_{\text{eNB-UEout,type1}}$$

In this simple setting, the overall relay system performance is the harmonic average of channel capacities of the backhaul and the access link, plus the channel capacity of donor eNB to the macro UE connection. The data rate of backhaul channel $R_{\text{eNB-RN}}$ plays crucial rule here since the access link rate may not be very high due to the following reasons, even though UE1 is quite closer to RN than donor eNB:

- Much lower transmit power of RN, lower antenna gain compared to donor eNB.
- Strong interference from donor eNB transmission which serves UE2.

Note that the second term ($R_{\text{eNB-UEout, type 1}}$) also suffers from the intra-cell interference caused by "type 1" RN transmissions.

Downlink performances of type 1 relay systems were evaluated by a number of companies. There are two major ways to place relay nodes within a macro network: random uniformly distributed, or near the edge.

3.4.4 Downlink Performance Evaluation with Uniformly Distributed Relay Nodes

With uniformly distributed UEs, the simulations show that roughly 40 % of UEs were attached to the RNs in Case 1 and 25 % of UEs attached to the RN for Case 3 for 4 RNs/cell case [8].

The different alternatives for relay placement are compared along with the RN antenna configuration optimization:

- Baseline, no site planning (B = 0, N = 1),
- Bonus SNR due to site planning (B = 5 dB, N = 1),
- Higher probability of LOS due to site planning (B = 0, N = 3), and
- Bonus SNR and higher probability of LOS (B = 5 dB, N = 3, for illustration only).

Two antenna configurations were considered for the backhaul reception at the RN:

- Omni-directional antennas without vertical pattern.
- Directional antennas pointing toward donor cell without vertical pattern.

As Fig. 3.18 shows, implementing directional antennas at RN backhaul link can improve SINR significantly compared to omni-directional antennas. It is also observed that for different ways of relay site optimization, the bonus SINR approach outperforms N = 3 higher probability of LOS approach in the case of uniformly randomly dropped of RNs.

Three situations were simulated in [8]:

- Out-of-Band relays
  - The backhaul bandwidth is unlimited. The backhaul does not consume the radio resources on the eNB-UE or RN-UE links.
- InBand relays with ideal backhaul link
  - The backhaul link (eNB-RN) shares radio resources with the eNB-UE and RN-UE links. The time domain resource partition is static, i.e., if x backhaul subframes of a radio frame are allocated to backhaul (10−x), subframes would be used to serve the macro UEs by the eNBs. Similarly (10−x), subframes are to be used to serve the relay UEs by RNs.
  - The backhaul spectral efficiency is assumed very high, so that the UEs served by the RNs would not experience any traffic bottleneck on the backhaul.
- InBand Relays with non-ideal backhaul link
  - The time domain resource partition between the backhaul link and access link, and the direct link is static. The backhaul spectral efficiency is assumed to be non-ideal where the UEs served by the RNs would sometimes experience traffic bottleneck on the backhaul.

Relay system results are shown in Figs. 3.19 and 3.20, for 4 RNs per cell and 10 RNs per cell, respectively. In both settings, there are total 25 UEs per cell.

For Case 1, the following performance characteristics are observed:

- The outband relays exhibit the highest gains for both the cell average and cell edge rate.
- Gains of the inband relays with non-ideal backhaul provide are very modest compared to the throughput of the baseline case without relay. For Case 1, the sector throughput and cell-edge throughput gains are 3 and 6 %, respectively when there are 4 RNs per cell.

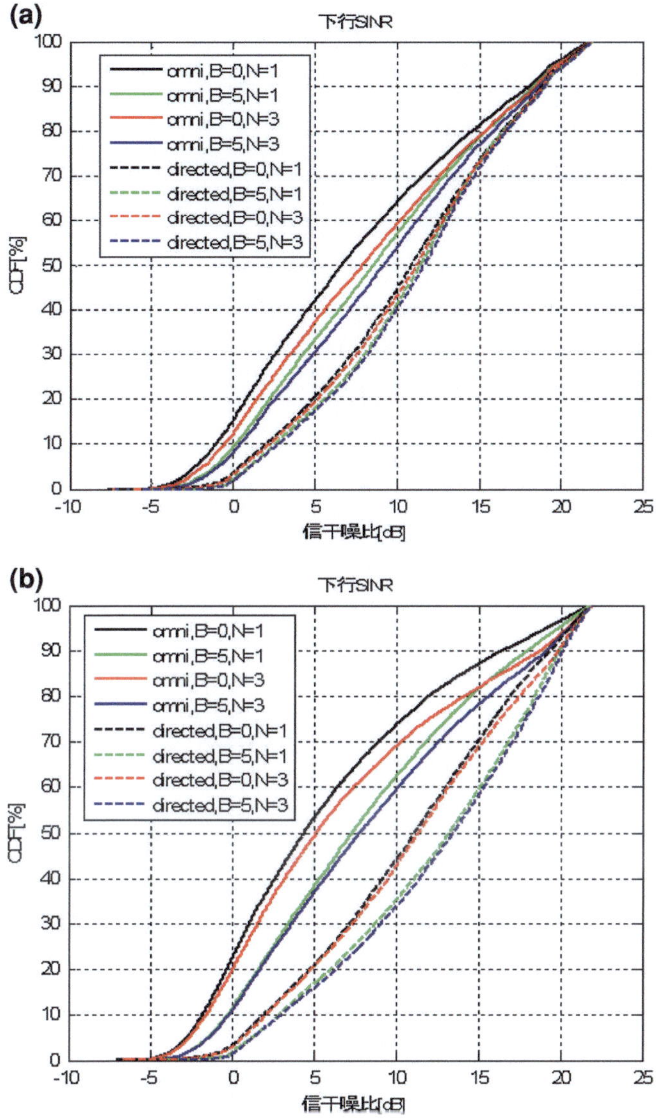

Fig. 3.18 CDF plots of backhaul C/I with different configurations. **a** Case 1, **b** case 3

- For inband relays with non-ideal backhaul, denser relay deployment (from 4 to 10) does not notably improve the performance. The relay cell throughputs are limited by the backhaul link throughputs and cannot be improved if the number of backhaul subframes is unchanged. It is noted that adding more relays in a small-cell system (e.g. Case 1 with ISD 500 m) is not always preferred as it increases the interference and can degrade throughput.

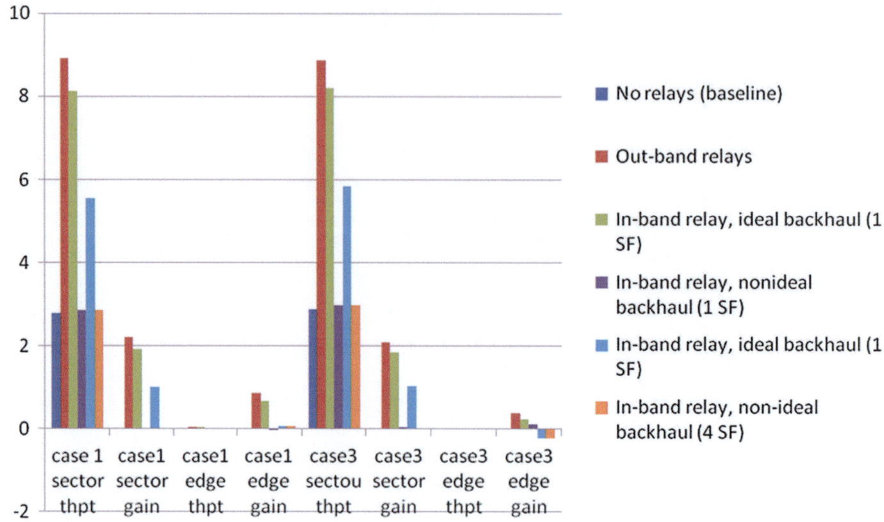

Fig. 3.19 Spectral efficiency gains in percentage, 4 RNs and 25 UEs per cell

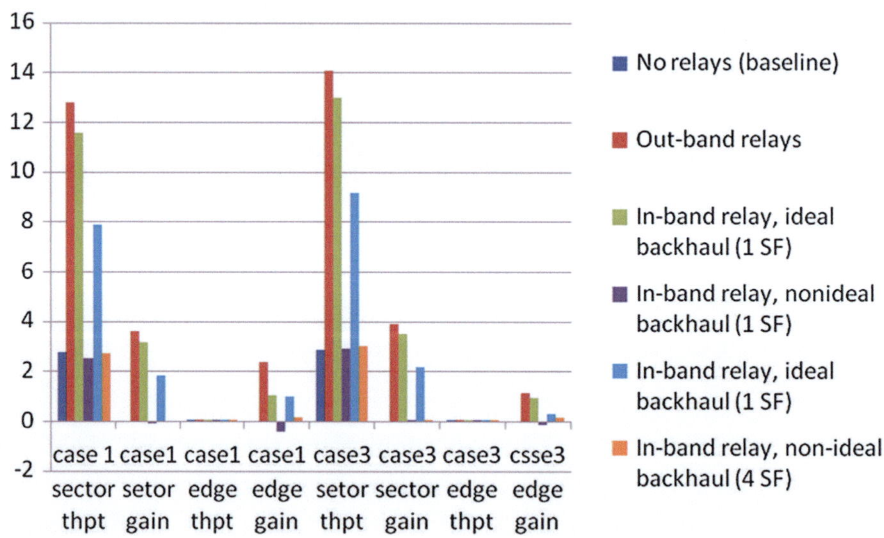

Fig. 3.20 Spectral efficiency gains in percentage, 10 RNs and 25 UEs per cell

- The performance of inband relays with ideal backhaul link sits between the out-of-band results and inband with non-ideal backhaul link. The sector throughputs roughly scale with the fraction of subframes available for access links, i.e., 9/10 for 1 backhaul subframe per radio frame, and 6/10 for 4 backhaul subframe per radio frame.

For Case 3, the followings are observed:

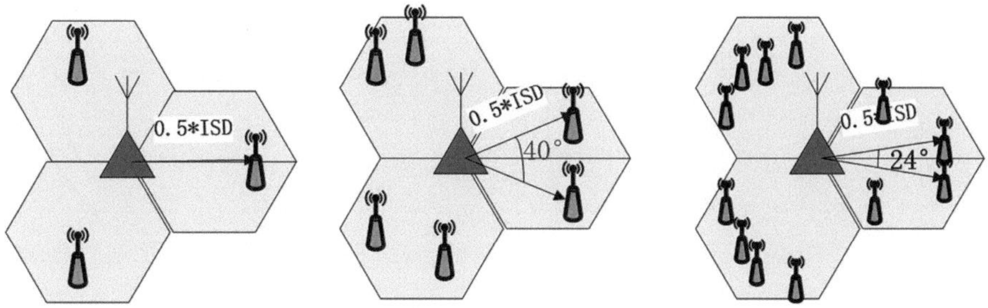

Fig. 3.21 Relay deployment patterns when relay nodes are placed near cell edges

- The out of band relays provide significant gains for both the sector throughput.
- For inband relays, better performance is observed in four RNs/cell assigned with 1 backhaul subframe, compared to 4 backhaul subframes. The opposite trend is observed for 10 RNs/cell.
- The performance of inband relays with non-ideal backhaul is quite poor.
- The sector throughput of inband relays with ideal backhaul link, as expected, the results scale with the fraction of subframes available for access links.

Note here a fixed and a simple access/backhaul resource split is assumed. Improved results are expected when optimum resource partitioning schemes are enabled. The results in [8] represent the lower bound of relay performance, due to the random locations of RNs and fixed backhaul allocation, i.e., no co-scheduling of macro UEs.

3.4.5 Downlink Performance Evaluation with Relay Nodes Placed Near Cell Edges

The position of the RNs plays major role in relay system performance. The RN location shall be carefully chosen, such that it can effectively provide coverage gain for cell edge UEs and at the same time have sufficient quality in the backhaul link. Moreover, the distance among RNs shall be large enough to avoid significant interference. In [9], the deployment of RNs is shown in Fig. 3.21. The RNs are evenly placed on the circle centered at eNB with a radius of 1/2 ISD.

The backhaul link geometry distributions are shown in Fig. 3.22 which behaves significantly different from those of randomly distributed RNs.

Except for the case of 4 RNs per cell, most RNs, i.e., >90 %, enjoy better than 10 dB average SINR. The SINR is capped at round 12 dB in Case 1 due to the significant inter-cell interference. Such cap almost disappears in Case 3 when the SINR can go up to 18 dB. The bumps observed in Fig. 3.22 reveal two groups of RNs: those in the middle of the bore-sight of eNB antenna, versus those on the outer side of the two RNs in the middle. Apparently, RNs of the first group receive stronger signal from donor eNBs and less interference from neighboring eNBs. Therefore,

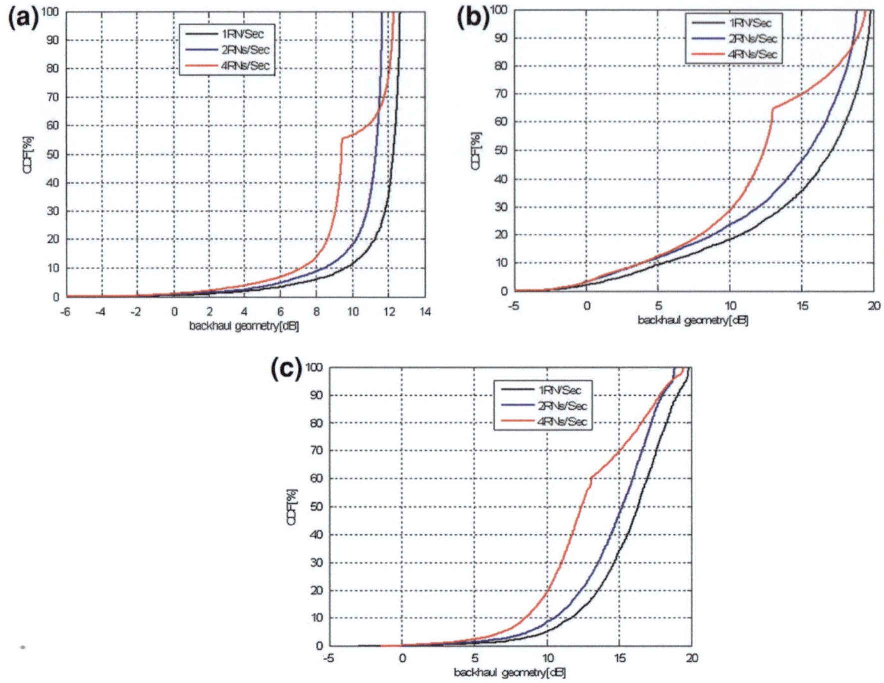

Fig. 3.22 Backhaul link geometry CDFs for **a** case 1, **b** case 3 suburban, and **c** case 3 rural

their SINR CDF is mainly reflected in the upper part of the curve in Fig. 3.22. The lower part is of RNs of the other group. Such geometry CDF discontinuity is less pronounced in Case 3 due to the lower interference from other cells.

The direct link (eNB-UE) and access link (RN-UE) geometry CDFs are shown in Fig. 3.23.

It is seen that as the number of RNs increases, more bad geometry UEs previously connected to eNB are now switched to RNs, leaving more good geometry UEs served by the donor eNB. Therefore, geometry CDFs of the direct link are improved. Pushing more UEs to RNs would degrade the overall geometry CDFs of RN cells, as some UEs in poor locations cannot get their SINR improved either by tagging on the donor eNB or one of the RNs. Hence, CDFs of geometry of access link are degraded as the density of RN increases. This trend is similar between in Case 1 and Case 3, except that access link geometry is always better than that that of direct link in Case 3, which can be explained by the less other cell interference. In another word, increasing the number of RNs is always preferable in power limited scenarios.

Cell selection for UE-RN association and UE-eNB association is based on RSRP without any bias. Figure 3.24 shows the percentage of UEs served by RNs (i.e., R-UEs) for different simulation cases.

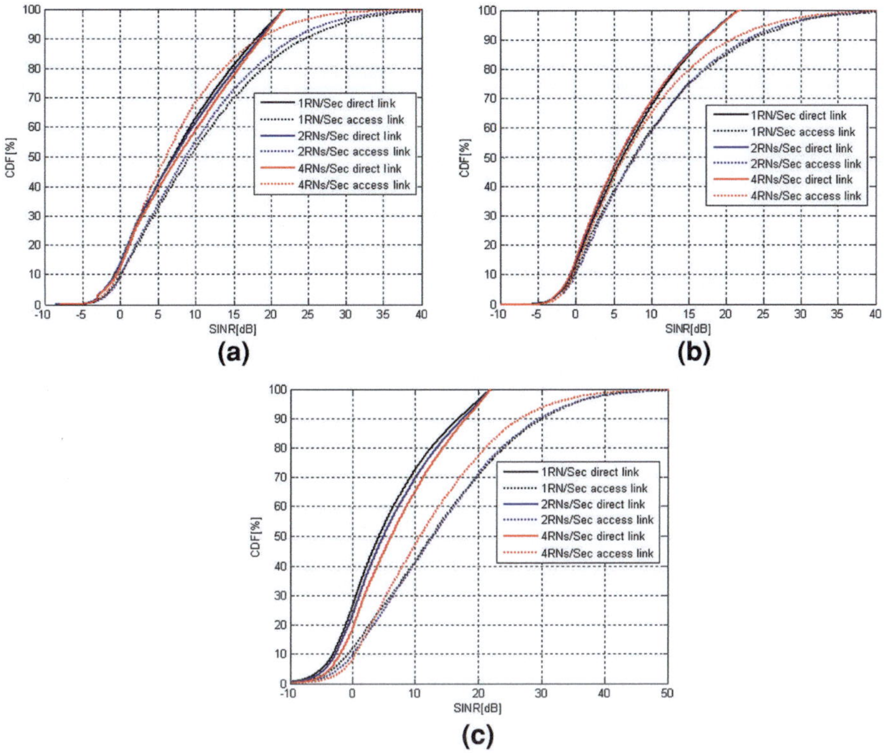

Fig. 3.23 Direct and access link geometry CDFs for **a** case 1, **b** case 3 suburban, and **c** case 3 rural

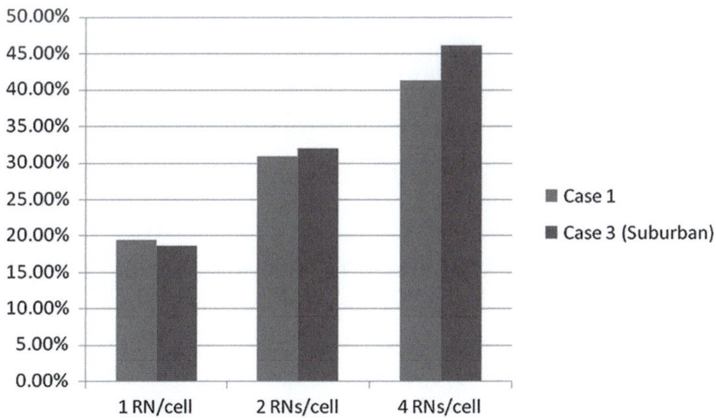

Fig. 3.24 Percentage of relay served UEs in case 1 (typical urban) and case 3 (suburban)

In the simulation study in [9], in-band half duplex is assumed. The resources in a backhaul DL subframe are exclusively used by DL transmission for RNs. In other DL subframes, eNB and RNs schedule transmissions for their serving

Table 3.5 Throughput results for case 1 with backhaul overhead

	BS thpt (bps/Hz)	RN thpt (bps/Hz)	Sum thpt (bps/Hz)	SUM thpt gain (%)	Cell edge thpt (bps/Hz)	Cell edge thpt gain (%)
No relay	1.9521	/	1.9521	0	0.0326	0
1 RN/cell	1.0869	1.0105	2.0973	7.4	0.0238	−27
2 RNs/cell	1.1398	0.9631	3.0660	57	0.0313	−4
4 RNs/cell	1.2170	0.7545	4.2350	117	0.0394	21

Table 3.6 Throughput results for case 3 suburban with backhaul overhead

	BS thpt (bps/Hz)	RN thpt (bps/Hz)	Sum thpt (bps/Hz)	Sum thpt gain (%)	Cell edge thpt (bps/Hz)	Cell edge thpt gain (%)
No Relay	1.6067	/	1.6067	0	0.0152	0
1 RN/cell	0.9188	1.3067	2.2255	38.5	0.0121	−20.4
2 RNs/cell	1.0680	1.1935	3.4552	115	0.0207	36
4 RNs/cell	1.1272	0.9580	4.9595	209	0.0330	117

UEs, i.e., macro UEs and R-UEs, respectively, using channel dependent proportional fairness scheduler.

Relay system throughput results are presented in Tables 3.7 and 3.8 where the backhaul transmission overhead is counted in. Note that the backhaul transmission is not explicitly modeled in simulations. The amount of required backhaul resources is estimated using the spectral efficiency obtained from Fig. 3.23, or more specifically assuming the mean backhaul geometry value. From this aspect, the results in Tables 3.5 and 3.6 only represent the approximate estimation of the relay system performance.

From the results it is seen that even when the backhaul overhead is considered, the gains from relay deployment are still significant in terms of sum capacity, as long as RN relay deployment reaches certain density, i.e., 4 RNs/cell and they are deployed near cell edges in a circle of a ring. When the density of RN is low, e.g. 1 or 2 RNs per cell, the relay system performance would be punished, compared to no relay case. In general, the gains are more significant in Case 3 where less other cell interference is seen.

It should be noted that the results in [9] represent the upper bound of relay system performance, considering that

- Near near cell-edge deployment of RNs. While it is desired to put RNs in poor geometry areas, real world cell topology is much more complicated and of less regular shape as in the simulations. Also the zone restriction may prohibit the exact RN planning that is purely based on clean geometry. So the deployment would be between the totally random and the pure geometrical plot on the blue print.
- Backhaul data buffering is not modeled, which would reveal the backhaul limitation. As seen in Sect. 3.4.4, such limitation would significantly reduce the performance potential of inband relay.

3.4.6 Uplink Performance Evaluation with Relay Nodes Placed Near Cell Edges

Similar to the RN layout in [9], relays in the uplink system evaluations in [10] are also placed at the cell edge to improve the poor geometry UEs. Denser deployment of RNs is tried, with the layout illustrated in Fig. 3.25. They resemble two rings. But the distances to the eNB are not exactly the same. For example, on the inner circle, the RN at the antenna boresight is the furthest, i.e., 0.415 of inter-site distance (ISD), followed by the two next to it with 0.395 ISD, and followed by the two on the side with 0.38 ISD. Similarly, on the outer circle, the furthest is at the boresight, 0.58 ISD to the eNB, followed by the two next to it, with 0.53 ISD, and followed by the two the side with 0.49 ISD.

In the backhaul subframe, RNs will transmit to DeNB and do not receive any data from relay-UEs. The uplink backhaul subframes are used exclusively for relay uplink transmission, i.e., no macro-UE transmits are scheduled in those subframes. The uplink backhaul subframe configuration is the same for all RNs in the simulation. At each RN, an omni-antenna is used for the access link reception and a directional antenna is used for the backhaul link transmission. 5 dB bonus is applied to the pathloss between the donor eNB and a relay node to reflect relay site planning gain. The HARQ round-trip time (RTT) for backhaul link is assumed to be 10 ms. The same power control procedure is used for both relays and macro-UEs. Only open loop power control is turned on. While it may not be the optimized setting for the network with both macro eNBs and relay nodes, for simplicity, the same fractional power control parameters (P0 and alpha) are used for both relay nodes and macro-UEs. Specifically, the following parameter settings are used. Case 1: P0 = −56 dBm; alpha = 0.6. Case 3: P0 = −67 dBm; alpha = 0.8

Antenna configurations and site-planning settings make huge difference in terms of backhaul link performance. In Fig. 3.26, four settings are compared. In Setting A, omni-directional antenna is used for backhaul link, without RN site planning. In Setting B, directional antenna is used for backhaul link, without RN site planning. In setting C, omni-directional antenna is used for backhaul link, with RN site planning. In Setting D, directional antenna is used in backhaul link, with RN site planning. The CDFs of wideband downlink SINR of backhaul link are compared in Fig. 3.26a and b for Case 1 and Case 3, respectively.

It is seen that about 10 dB can be gained for backhaul SINR by using directional antenna between relay and DeNB for both Case 1 and Case 3. The site planning could further improve the backhaul quality by about 6 dB for Case 1 and about 1.7 dB for Case 3. The site planning gain is smaller in Case 3 than in Case 1. It can be explained by the fact that in Case 3, the interference is mainly from two other sectors of the same donor eNB. In this power limited scenario, adding 5 dB as pathloss bonus to eNB-RN connection does not benefit backhaul link SINR. For system performance evaluation, Setting D is used for backhaul link.

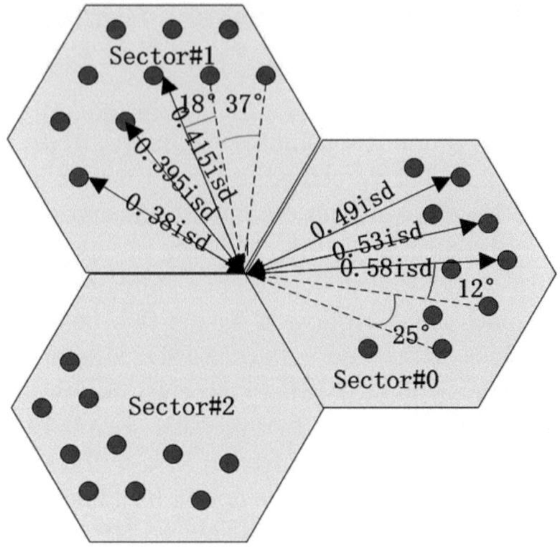

Fig. 3.25 Two-ring relay deployment with 10 RNs/cell

Simulations in [10] assume proportional fairness scheduling at both eNBs and RNs. The backhaul link is assumed non-ideal, i.e., its capacity is limited and would throttle the data relaying from RN served UEs to the donor eNB. The buffer size at the RN is set based on capacity of the backhaul uplink of that RN. Frequency reuse factor is 1, i.e., macro UEs and RN served UEs can fully reuse the uplink resource.

The control channel overhead is 2 PRBs over 10 MHz operating bandwidth, i.e., 48 PRBs are used for uplink data transmission. In the absence of agreed fast fading model for relay simulations, TU channel is used for direct link, access link and backhaul link. UE-RN and UE-eNB association is based on RSRP. Note that the transmission buffer at RN may not be full, although UEs served by the RNs are full buffer.

The simulations consider 1, 4, and 10 relay nodes per sector. The time domain resources allocated to the backhaul subframes can be 1, 2, 4, 6 per radio frame. It is expected that as more subframes are configured for the uplink backhaul, the throughput of macro cell (or equivalently for macro UEs) would decrease since the backhaul resource is solely for RN-eNB communications.

The relay system uplink throughput corresponding gains are shown in Figs. 3.27 and 3.28, for Case 1 (urban) and Case 3 (suburban), respectively. It is observed that for each RN density (whether 1, 4 or 10 RNs/cell), there is an optimal ratio of backhaul subframe (1, 2, 4, or 6 per radio frame) that can bring the highest throughput gain. For example, in Case 1 with one RN per cell, 1 backhaul subframe per radio frame configuration can improve the average throughput by about 5 %, whereas allocating 4 or more backhaul subframes would degrade both the cell average throughput and edge throughput. The optimal value of number of backhaul subframes seems increasing as more RNs are deployed in a cell. The absolute throughput gains are also increased with more RNs. For example, as the RN density goes to 10, the average throughput gain can reach 35 % and the edge throughput gain can reach 79 % in Case 1.

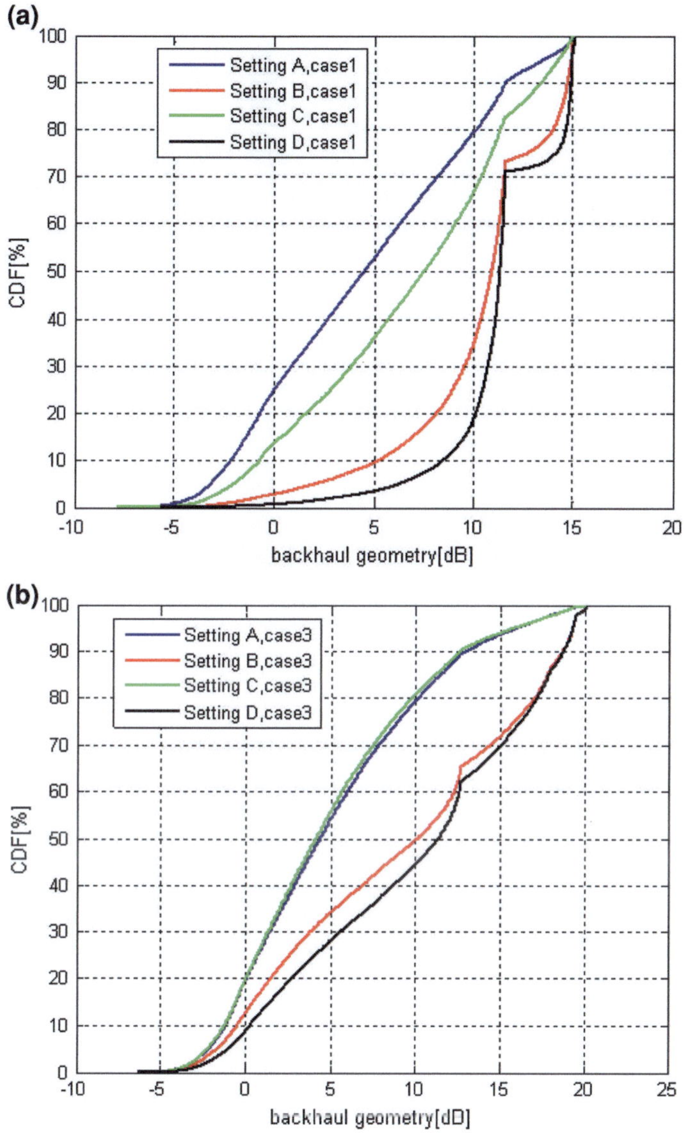

Fig. 3.26 CDFs of backhaul wideband DL SINR under different settings. **a** Case 1, **b** case 3

The uplink gains shown are higher than the relay gains in downlink simulations in general. The significant improvement in the average UE link quality for both access link and direct link would be a main reason. Poor geometry UEs previously had to transmit high power to reach eNB. Now they are offloaded to the nearest RNs and can transmit with lower power. Therefore, uplink interference is reduced.

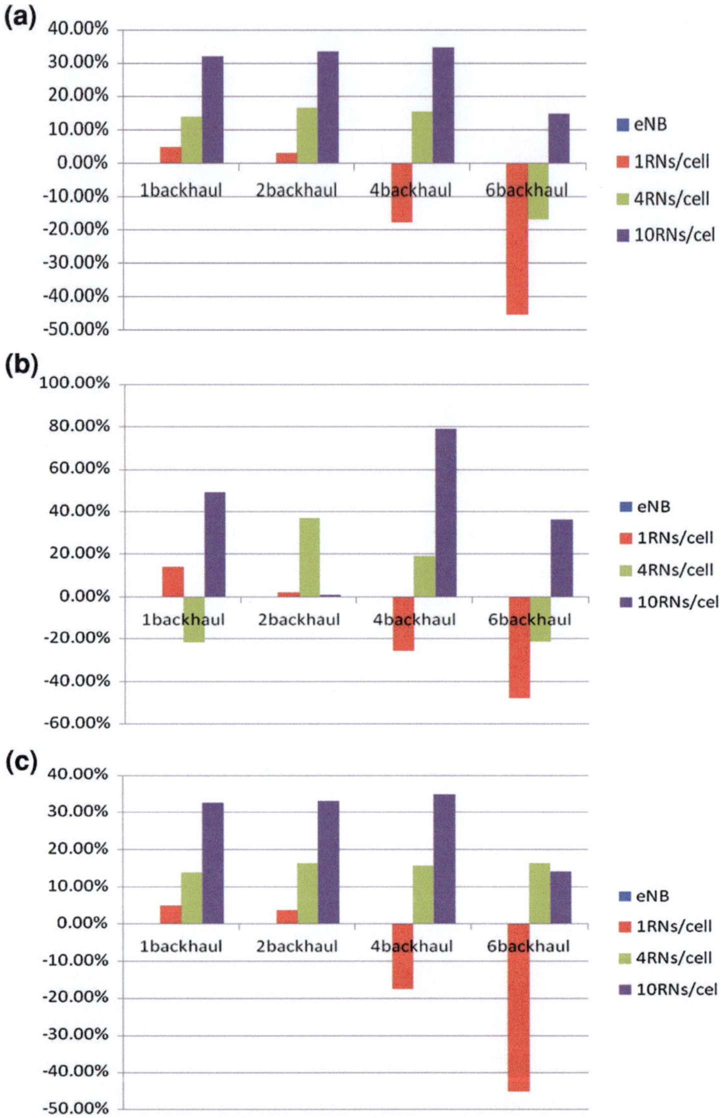

Fig. 3.27 Relay uplink throughput gains for case 1. **a** Average aggregate throughput gain, **b** 5 %-tile UE throughput gain, **c** average UE throughput gain

The uplink throughput gains for Case 3 are shown in Fig. 3.28. Similar to Case 1, the number of backhaul subframes needs to be increased when more RNs are deployed, in order to avoid the capacity bottleneck in backhaul link.

A major observation is that the relay gains in Case 3 are much higher than in Case 1. For example, with 4 RNs/cell and 2 backhaul subframes per radio frame, the average throughput gain can reach 54 % and the cell edge throughput gain is

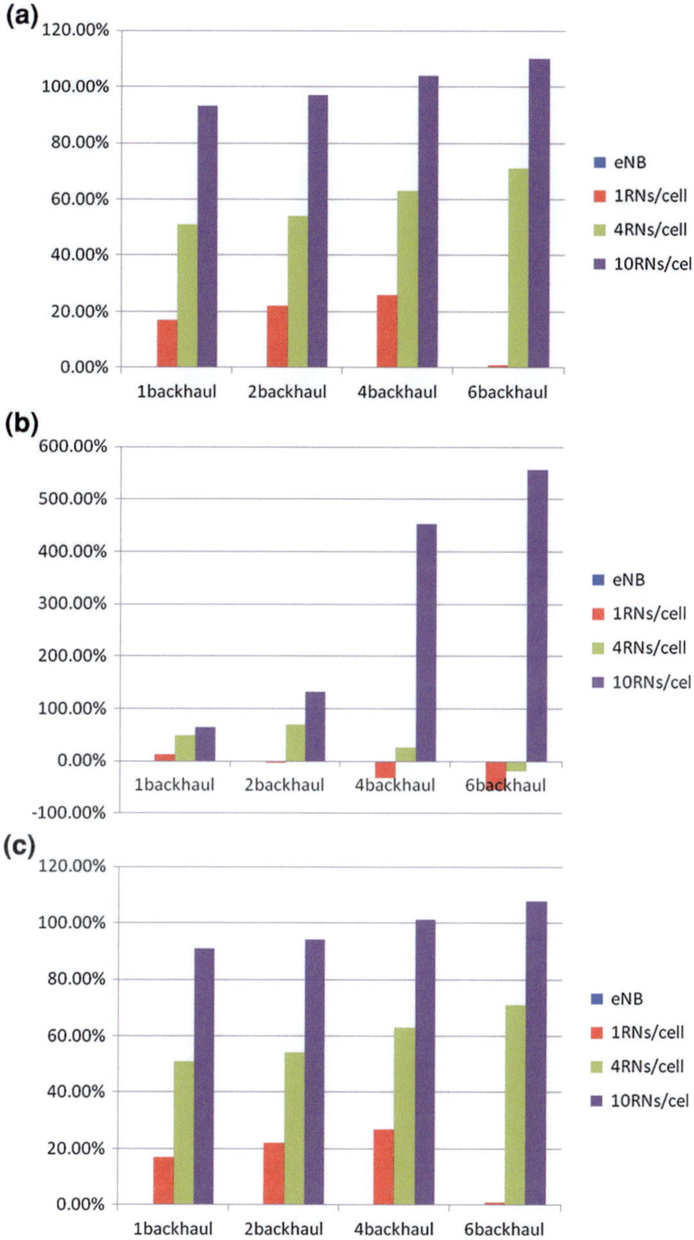

Fig. 3.28 Relay uplink throughput gains for case 3. **a** Average aggregate throughput gain, **b** 5 %-tile UE throughput gain, **c** average UE throughput gain

about 70 %. With 10 RNs/cell and 6 backhaul subframes, cell average throughput gain can go up to 110 % and the cell edge gain reach nearly 560 %. The gain is mainly contributed by the RN cell where relay served UEs experience much less interference and very high SINR compared to Case 1.

3.5 Type 2 Relay

The motivation of type 2 relay is to exploit the cooperative nature of relay. In type 1 relay, the first hop transmission in downlink targets only for the relay node, i.e., not useful at all to relay served UEs, even though it contains the information bits ultimately for those UEs. During the second hop, the donor eNB and the type 1 RN schedule their own UEs, without any cooperation in general. In contrast, type 2 relay would make use of the first hop transmission for its served UEs, and facilitate cooperative transmission in the second hop.

3.5.1 Definition

A "type 2" relay node is an inband relaying node characterized by the following [1]:

- It does not have a separate Physical Cell ID and thus would not create any new cells
- It is transparent to Rel-8 UEs; a Rel-8 UE is not aware of the presence of a type 2 relay node
- It can transmit PDSCH
- At least, it does not transmit CRS and PDCCH

Type 2 relay assumes centralized resource scheduling by donor eNB. That is: eNB sends explicit control signaling to type 2 RN and indicates the transmission format for access link. Demodulation reference signal (DRS) is needed for data demodulation in the access link.

Since type 2 relay does not have cell ID, it would not have any RRC functionalities and cannot support mobility management of UEs. Hence, it looks closer to L2 relay. However, type 2 relay itself does not have scheduler, i.e., the resource scheduling is done by the donor eNB. In this sense, type 2 relay is a mix of L1 relay and L2 relay. Note that, certain MAC layer functionalities would be needed in type 2 relay to fully take advantage of cooperative relaying. So it leans more towards L2 relay.

3.5.2 Technologies

Several transmission schemes of type 2 relay were discussed in [11]: cooperative transmission, resource reuse, and pure overhearing.

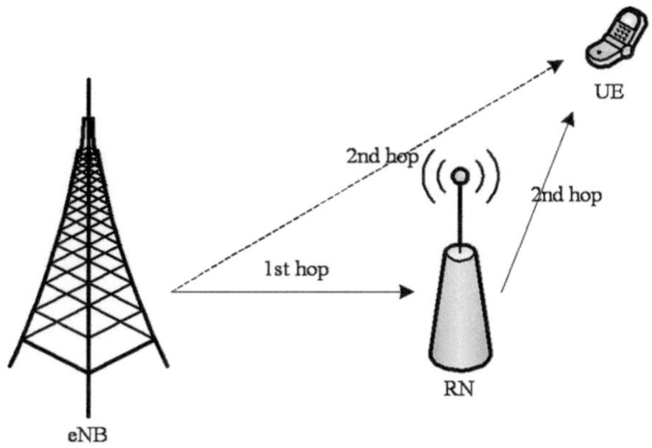

Fig. 3.29 Cooperative transmission of type 2 relay

Cooperative transmission mode is suitable in scenarios where donor eNB and relay coverage are overlapped. As illustrated in Fig. 3.29, DeNB sends downlink transmission block (DL TB) and DCI 2 of target R-UE to RN, while DCI 2 indicates the scheduling information for access link in the second hop. The backhaul transmission uses HARQ process, and only when DeNB receives ACK from RN, the initial transmission in the second hop is based on the assigned resources by DCI 2. For the second hop transmission, DCI 2 is sent by DeNB and TB is transmitted by both DeNB and RN. And then, RN overhears the ACK/NACK from R-UE, in the case of NACK, both DeNB and RN would carry out the retransmission.

For HARQ retransmissions, there are two alternatives:

- Synchronous and non-adaptive retransmission.

Using synchronous and non-adaptive retransmission for access link has some advantages such as no extra control signal needed for scheduling retransmissions, and no delay introduced for HARQ process. After 4 subframes, non-adaptive retransmission is carried out by DeNB and RN simultaneously.

- Asynchronous and adaptive retransmission

Adaptive retransmission can provide more flexibility and channel adaptability. For the sake of higher retransmission efficiency on access link through dynamically scheduling, some re-assignment information needs to be communicated between DeNB and RN on backhaul. After receiving new DCI 2, RN and DeNB retransmit to R-UE on assigned resources.

The DL HARQ process timing of scheme 1 is showed in Fig. 3.30:

k1—4 ms is the preferred value for k1, which means that backhaul DL transmission occurs in subframe n and UL transmission occurs in subframe n + 4. Such setting facilitates the subframe allocation between backhaul link and access link, avoids the possible collision between RN UL backhaul transmission and overhearing of R-UE HARQ feedback. k1 can take some other values depending on the capability of RN.

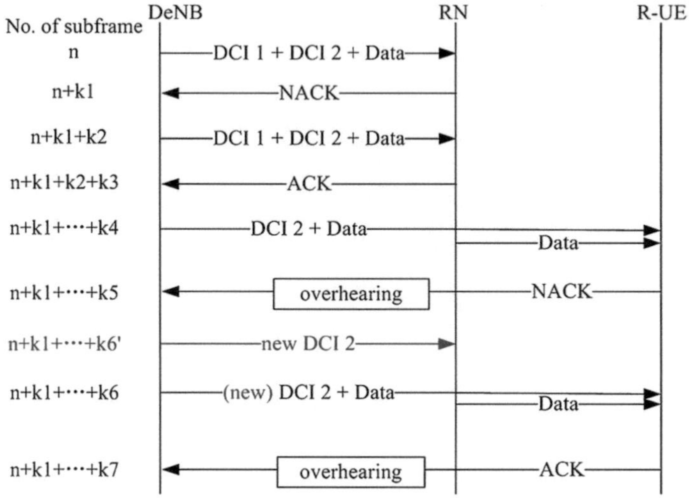

Fig. 3.30 HARQ process of cooperative transmission for type 2 relay

k2—It depends on the processing (demodulation, decoding, scheduling) capability of DeNB and the allocation of backhaul link DL subframe. In other words, $k_{2\min} = \min\{4\,\text{ms}, N_{next\,\text{backhaulsubframe}}\}$.
k3—the same as k1.
k4—to decide the value of k4, two factors should be considered. On one hand, RN needs implement re-coding/modulation/mapping to assigned resource of the correctly received TB in k4 ms completely. On the other hand, while NACK is received, DeNB should release and reschedule the resources assigned by DCI 2 for some other M-UEs during k4 ms.
k5—according to Rel-8 protocol, $k_5 = 4\,\text{ms}$.
k6′—when asynchronous adaptive HARQ retransmission is used on backhaul link, shown in blue in Fig. 3.30, DeNB needs to indicate RN the new DCI 2. And then according to new DCI 2, RN cooperates with DeNB in the retransmission to R-UE. To indicate the new DCI 2, a backhaul DL transmission should be assigned.
k6—when synchronous non-adaptive HARQ retransmission is used on backhaul link, $k_6 = 4\,\text{ms}$, no backhaul control information is communicated between DeNB and RN, the same assignment, such as MCS and resources is used for retransmission.
k7—according to Rel-8 protocol, $k_7 = 4\,\text{ms}$

Resource reuse mode is suitable when coverage areas of RNs are not overlapped [11] as shown in Fig. 3.31. Therefore, frequency resource can be reused by multiple RNs in the second hop to transmit different data to their served UEs. The DL HARQ procedure of scheme 2 is the same with cooperative transmission scheme in Fig. 3.30, retransmission may use synchronous and non-adaptive mode or asynchronous and adaptive mode, the differences between the two schemes is that on second hop DeNB only send DCI 2.

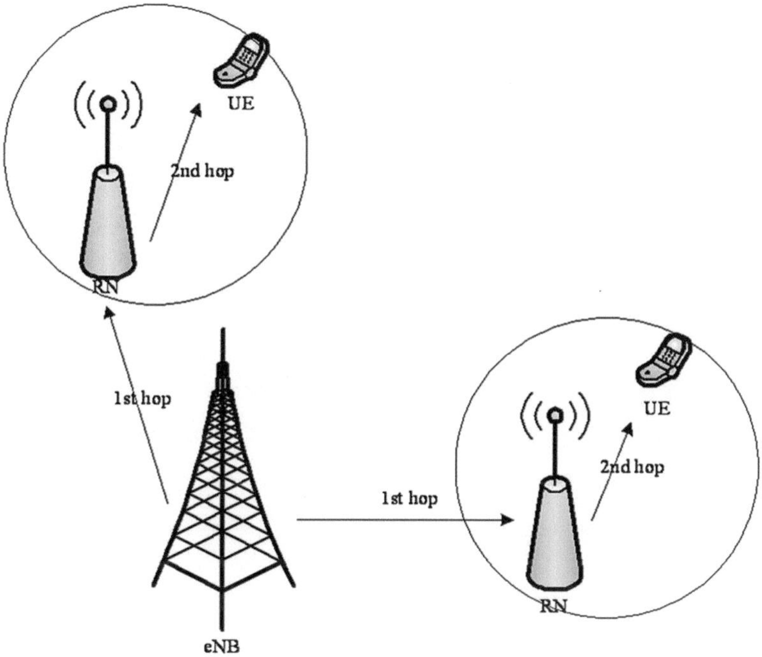

Fig. 3.31 Downlink cooperative transmission and resource reuse for type 2 relay

A simple transmission scheme of Type 2 relay is that RN overhears the direct link initial transmission and R-UE feedback and assists retransmission if NACK is received. In case of transport block is not decoded correctly by RN, retransmission would be carried out only by DeNB.

Pure overhearing scheme has the least need for control information communication on backhaul link as seen in Fig. 3.32. After some initial pairing of RN and R-UEs, for which R-UE a RN should overhear and provide service, no more control information is needed to indicate RN. When direct link initial transmission from DeNB is not correctly received by target R-UE, yet RN decodes the transport block correctly, and then DeNB and RN should implement a synchronous non-adaptive retransmission 4 ms later after NACK. In case of RN does not receive the TB correctly, no signal is transmitted on access link.

The drawback of pure overhearing scheme is that the first transmission does not consider channel quality in the backhaul link (eNB-RN) which is typically much better than the direct link (eNB-UE). The MCS based on the direct link measurement would be too conservative for the backhaul link, making the two-hop relaying sometimes even less efficient than direct link.

Besides assisting downlink transmission, type 2 relay can also improve the efficiency of uplink transmission. Synchronous HARQ process is adopted in LTE UL transmission. So it is preferable that UL transmission scheme of Type 2 relay follow a strict timing in order to fulfill backward compatibility to Rel 8 UEs.

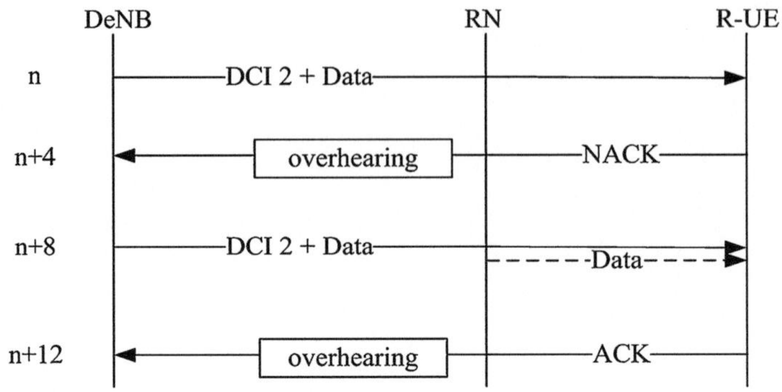

Fig. 3.32 HARQ process of pure overhearing for type 2 relay

Considering of the transparent character of Type 2 relay node, two UL trans-

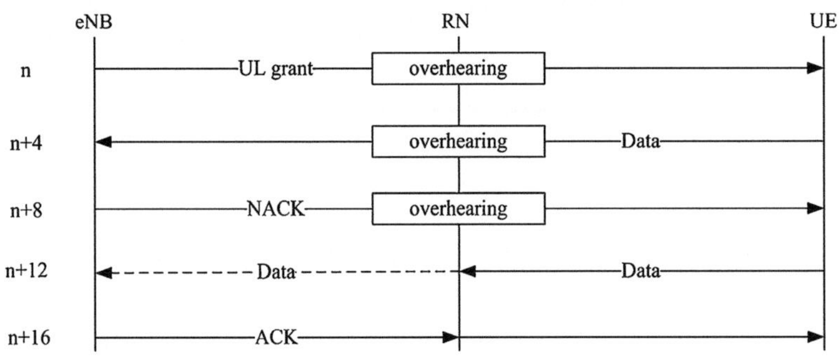

Fig. 3.33 HARQ process of UL transmission scheme 1

mission schemes and corresponding HARQ procedures can be considered:

Scheme 1: *overhear and assist retransmission*
Scheme 2: *RN feedback and retransmit on backhaul link*

- Scheme 1

RN overhears PDCCH of DeNB and gets the UL Grant for target R-UE, and then overhears the R-UE UL transmission. If DeNB receives the transport block correctly, it sends ACK via PHICH. If NACK is received, non-adaptive retransmission is carried out by both R-UE and RN, shown in Fig. 3.33:

Due to the synchronous HARQ on uplink, this is not enough time for the backhaul to communicate the control information of retransmission, such as resources assignment, new MCS, etc. Therefore, only non-adaptive retransmission can be carried out.

In case of the transport block is not overheard correctly, RN cannot assist the retransmission, and no combination gain can be achieved.

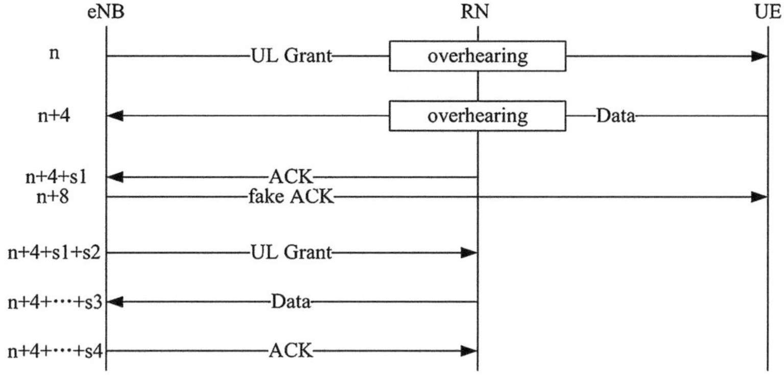

Fig. 3.34 HARQ process of UL transmission scheme 2 when the access link transmission succeeds

Because of the synchronous nature of UL HARQ process, R-UEs served by the same RN should be carefully scheduled in order to avoid the collision of RN between backhaul uplink transmission and overhearing on access uplink. It brings some restriction on the scheduling of R-UE.

- Scheme 2

RN overhears PDCCH, gets UL Grant and R-UE uplink transmission, and then RN feeds back to DeNB. Based on the feedback and the outcome of the transport block decoding at DeNB, retransmission may be assigned for backhaul link or direct link.

Figure 3.34 shows a HARQ procedure when the initial uplink transmission fails on subframe n + 4 + s1. RN sends its feedback of the decoding outcome, and then DeNB assigns RN to retransmit the transport block on backhaul link, and may indicate a "fake ACK" to R-UE.

The corresponding parameters are:

s1—if RN has sufficient capability to provide the feedback before DeNB sends PHICH to R-UE, DeNB may decide to indicate R-UE ACK or NACK according to the feedback from RN. When the feedback is NACK, DeNB should assign and schedule a retransmission for R-UE.

If the feedback from RN cannot be received early enough, i.e., $s_1 \geq 4$ ms, DeNB would send an ACK to R-UE. Then DeNB assigns backhaul retransmission when RN correctly decodes the transport block and sends an ACK. If RN still cannot decode the transport block correctly, DeNB would schedule R-UE retransmission in the direct link.

In addition, the value of s1 is also restricted by the subframe allocation of backhaul and access link because RN cannot overhear access uplink and transmit on backhaul uplink simultaneously.

s2/s3/s4—indicating the timing for the scheduling, data transmission and feedback on backhaul link, respectively. They are determined only by the subframe

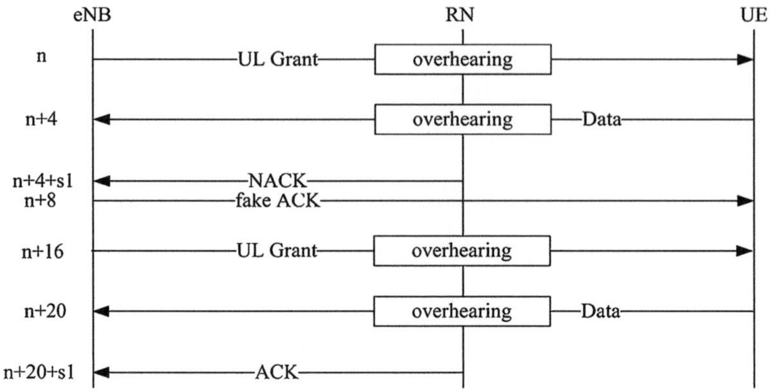

Fig. 3.35 HARQ process of UL transmission scheme 2 when the access link transmission fails

allocation of backhaul link. To minimize end-to-end delay, the value of s2, s3, s4 should be made as small as possible.

Figure 3.35 shows the condition when a transport block is not received correctly by the RN. A "fake ACK" is sent to R-UE, and then the DeNB schedules resources and indicates R-UE to implement a retransmission in direct uplink. According to synchronous HARQ procedure, the next retransmission may be assigned on subframe $n + 20$ when RN continues to overhear the access uplink transmission.

Base on the discussion above, scheme 2 may reduce the times for retransmission and the consumption of battery for R-UE in condition on successful decoding of the transport block at RN. Even though the feedback from RN is NACK, DeNB could reschedule R-UE retransmission of the transport block in direct link, with some delay. Comparing with scheme 1, some backhaul control information is needed to indicate the outcome of RN decoding.

In [12], a new error coding scheme was proposed to take advantage of cooperative transmission in type 2 relay. The coding protocol was termed "Time Division Multiple Access (TDMA)" relay code. The benchmark for comparison is the so called Multi-Hop (MH) based relay techniques, which is essentially type 1 relay. It is noted that existing 3GPP turbo-code can also be supported in such scenario of type 2 relay, with less performance benefit.

In "TDMA" scheme, the relaying has two stages for the transmission from the source to the destination, connected with relay as illustrated in Fig. 3.36. During the first stage, the source (the eNB) broadcasts a message to both relay and the destination (the UE). Therefore, this stage is also called Broadcast (BC) mode. During the second stage, the source (the eNB) remains silent, while the relay transmits to the destination terminal (the UE). The second stage is also called Multiple Access (MA) mode. Multiple access is a terminology in information theory, representing the case that the multiple sources of information available to

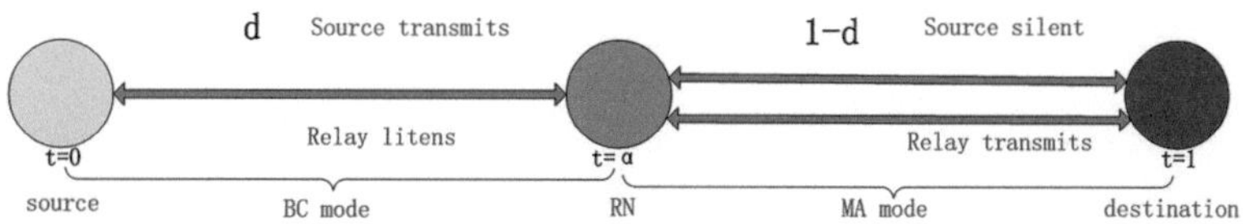

Fig. 3.36 "TDMA" relaying

improve the channel capacity. It is used here to emphasize the fact that the UE indeed has both data from the source in broadcast mode (first stage) and the data from the relay in the second stage that can be potentially utilized to boost the overall relay system capacity.

Below are some code design principles of the TDMA relay channel:

1. The eNB should targets a throughput (for a given resource block shared between source and relay) that is near the TDMA relay capacity.
2. The eNB transmits a broadcast mode codeword based on target throughput as described above. Note that the effective broadcast mode code rate (between eNB and RN) should approach the eNB-RN capacity, according to the TDMA capacity expression.
3. The RN sends rate-compatible parity bits (compatible with the broadcast mode received bits) during the multiple access mode.

Similar to the downlink cooperative transmission discussed earlier, CQI report for the backhaul link (eNB-RN) is needed to fully exploit the gain. A link level performance evaluation was conducted in [12] to compare the throughput potential of the TDMA relay. For simplicity, the RN is assumed to be on a unit-line between the source and destination as seen in Fig. 3.37. More specifically, RN is assumed to be half-way between the source and the destination, and the pathloss exponent is assumed 3. No shadowing fading or fast fading is considered.

Point-to-point communication, i.e., without relay, is also compared. To make the comparison fair, the same total power by the source is assumed in both point-to-point and TDMA relay. In another word, the total power is split between the first stage (broadcast mode) and the second stage (multiple access mode).

Assuming Gaussian alphabets, the link capacity results are shown in Fig. 3.37, compared with Multi-Hop (type 1 relay) and a repeater (amplify-and-forward relay). Results with finite alphabets are shown in Fig. 3.38. Significant gains of TDMA relaying over type 1 relay, repeater and no relay are observed, especially at medium and high SNR regions.

TDMA relaying fits well to HARQ mechanism in LTE as depicted in Fig. 3.39. Legacy turbo-codes in 3GPP can be re-used within the TDMA HARQ framework. It is noted that the overall spectral efficiency of the TDMA relay channel can be maximized by choosing more appropriate codes. Among the various candidate codes, low density parity check (LDPC) codes seem promising. There are more design freedoms in LDPC codes to construct and optimize the codewords to tailor specifically for typical relay geometry.

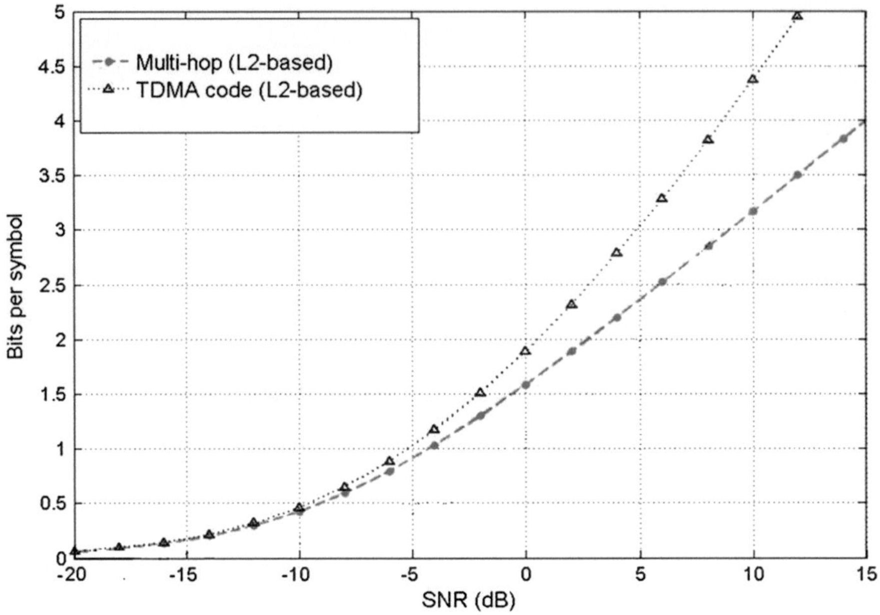

Fig. 3.37 Capacity of TDMA relay compared to type 1 relay, repeater and point-to-point transmission (no relay), assuming Gaussian alphabets

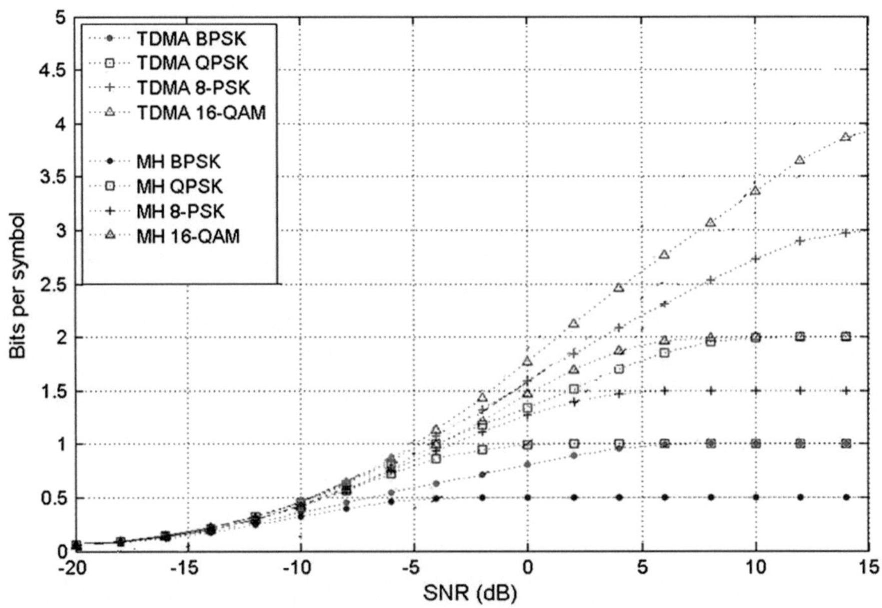

Fig. 3.38 Capacity of TDMA relay compared to type 1 relay, repeater and point-to-point transmission (no relay), assuming QAM alphabets

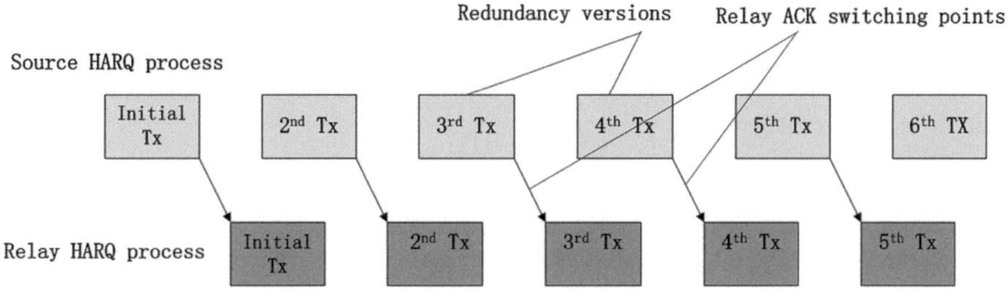

Fig. 3.39 HARQ process to support TDMA relaying

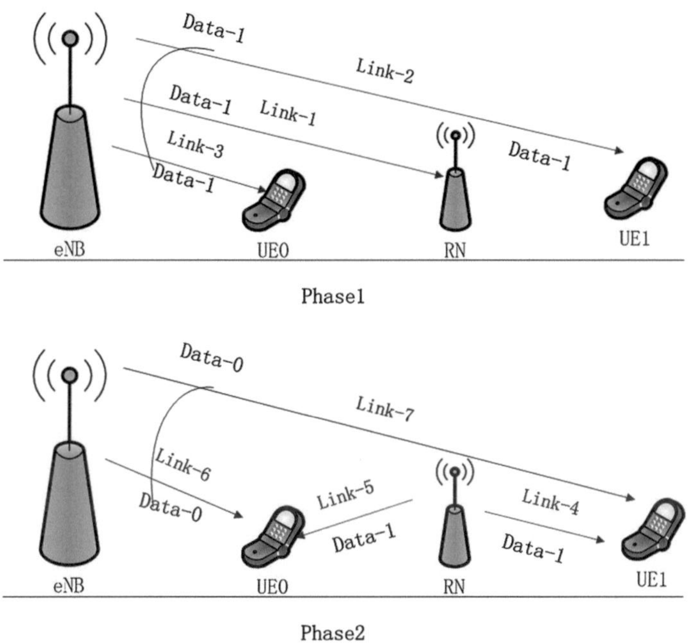

Fig. 3.40 A type 2 relay based on interference cancellation

The first hop transmission can be utilized in different ways, besides the code domain HARQ as in "TDMA" relay. In [13] a resource reuse scheme was proposed as shown in Fig. 3.40. This scheme relies on certain interference cancellation at the UE receiver. It can be categorized as a type 2 relay, even though the cooperation is done in a less direct way: through interference cancellation.

While traditionally considered as a pure product realization, receiver side interference cancellation has gradually become the mainstream implementation by mobile terminal vendors, especially for high-end chipset manufacturers. The study of cooperative relay, or more general cooperative transmissions, needs to consider the opportunities made available by such interference cancellation capability. In some sense, the process of interference estimation itself reveals more information at the source(s), which can be used to improve the overall capacities of all links that are involved.

3.5.3 Performance Evaluations

At high level, we can use the similar semi-analytical approach in Sect. 3.4.3 to estimate the system performance of type 2 relay. For example, for cooperative transmission mode, the channel capacity of the second hop can be written as

$$R_{\mathrm{RN-UEin,Type2}} = 1 + \log_2\left[\frac{P_{\mathrm{RN}}|h_{\mathrm{RN-UEin}}|^2}{I_{\mathrm{o,UEin}}}\right]$$

The overall link capacity of after two hops is:

$$R_{\mathrm{eff,Type2}} = \frac{R_{\mathrm{eNB-RN}}(R_{\mathrm{RN-UEin,L2}} + R_{\mathrm{eNB-UEin}})}{R_{\mathrm{eNB-RN}} + R_{\mathrm{RN-UEin,L2}}}$$

Due to the cooperative gain (appearing as the extra summation term $R_{\text{eNB-UEin}}$ in the numerator) and less interference from the macro, i.e., $R_{\text{RN-UEin, type1}} < R_{\text{RN-UEin, Type 2}}$), $R_{\text{eff, Type 2}}$ is significantly higher than the two-hop link capacity for UE1 in type 1 relay case. On the other hand, type 1 relay allows cell splitting gain, appearing as the term $R_{\text{eNB-UEout, type 1}}$ for UE 2 in Fig. 3.20. Numerical results would show whether the cooperative transmission in type 2 relay or cell splitting in type 1 relay can provide more performance benefit.

Type 2 relay performance was evaluated in [14]. Some simulation parameters follow TR 36.814. In particular, 10 RNs are placed in each cell and RN locations are uniformly distributed, i.e., uncorrelated. On average, there are 25 UEs in each cell and UEs are uniformly distributed over the network. The maximum transmit power of RN is 30 dBm.

Two receive antennas and single transmit antenna are assumed at RN. While high order MIMO, e.g., 4 × 4, can be used for backhaul link to improve the data rate, the propagation environment between eNB and RN is expected to have strong LOS component where the rank of the spatial channel may not be sufficient for the multiplexing, especially for 4 × 4 MIMO.

In this simulation, the constraints on MBSFN subframes for backhaul link are ignored in type 1 relay. Note that a type 1 relay creates a new cell and thus increases the total resource of the system, which brings potential gains in cell throughput. However, it is also noted that the user throughput served by type 1 relay can be significantly lower than the average user throughput served by the donor eNB. Hence, the user throughput fairness for all users in a cell (including both donor eNB served UEs and RN served

Table 3.7 Throughput gains, over-the-air combining for type 2 relay, no cross eNB and RN antenna precoding. RN randomly dropped

	Cell throughput gains			5 % user throughput gains		
	Type 1 RN (%)	Type 2 RN (single RN-eNB coop) (%)	Type 2 RN (multi RNs-eNB coop) (%)	Type 1 RN (%)	Type 2 RN (single RN-eNB coop) (%)	Type 2 RN (multi RNs-eNB coop) (%)
Case 1	−3.7	19.4	20.2	−24	15.1	17.5
Case 3	3.5	19.6	20.4	−76	10.8	14.7

UEs) could be suffered when RN served users are too frequently scheduled, which also introduces more interference to eNB served UEs. In the simulation, we adjust the scheduling frequency between macro-UEs and relay UEs, to maintain certain fairness criteria. Note that cooperative cell silencing is not assumed for type 1 relay.

Over-the-air combining is assumed for type 2 relay, without precoding among transmit antennas across donor eNB and RN. Two configurations of type 2 relay are considered: only the closest RN or multiple nearby RNs participate in cooperative transmissions with the direct link. We assume that reference signal, either CRS or DRS, is available for PDSCH demodulation. The overhead of DRS is not accounted.

As seen in Table 3.7, the average cell throughput and 5 % user throughput without relays are 11.86 Mbps and 126 kbps, respectively, for Case 1. In Case 3, the corresponding cell throughput and edge throughput are 10.92 Mbps and 102 kbps, respectively. Throughput gains are listed in Table 3.7. It is seen that for Case 1, type 1 relay causes slight loss in average cell throughput and moderate loss in cell edge throughput. For Case 3, type 1 relay provides slight gain in average cell throughput and significant loss in cell edge throughput.

In contrast, type 2 relay can improve average throughput by about 20 % in both Case 1 and Case 3. Cell edge throughput gain is about 15–17.5 % for Case 1 and 11–15 % for Case 3. The performance difference between single RN and multi-RN cooperation is not significant, except for cell edge throughput in larger cells, e.g., Case 3.

User throughput CDFs are shown in Figs. 3.41 and 3.42 for Case 1 and Case 3, respectively. The throughput fairness is significantly improved in type 2 relay, whereas certain fairness degradation is observed in type 1 relay, especially in Case 3.

In [15], a type 2 relay was put in a network to enhance the system performance when soft frequency reuse (SFR) is implemented. The UEs in the network are grouped into relay served UEs, macro UEs at cell edges and macro UEs close to the donor eNB.

The system operating bandwidth is divided into four subbands, e.g. F1, F2, F3 and F4, shown in Fig. 3.43. Subband F4 is allocated to macro UEs close to the eNB, with subbands F1, F2 and F3 assigned to different three sectors to avoid excessive other cell interference. Dynamic resource allocation can be achieved, as

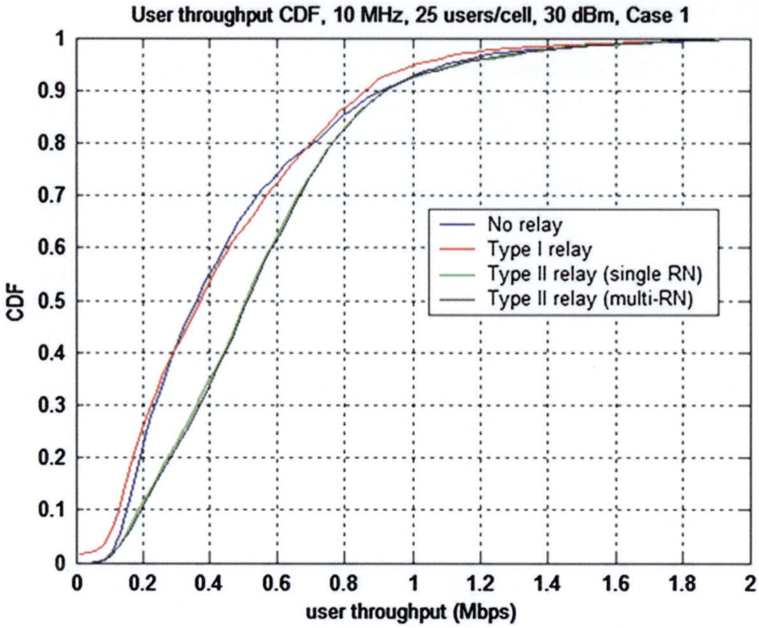

Fig. 3.41 Comparison of user throughput CDFs of different relays, case 1

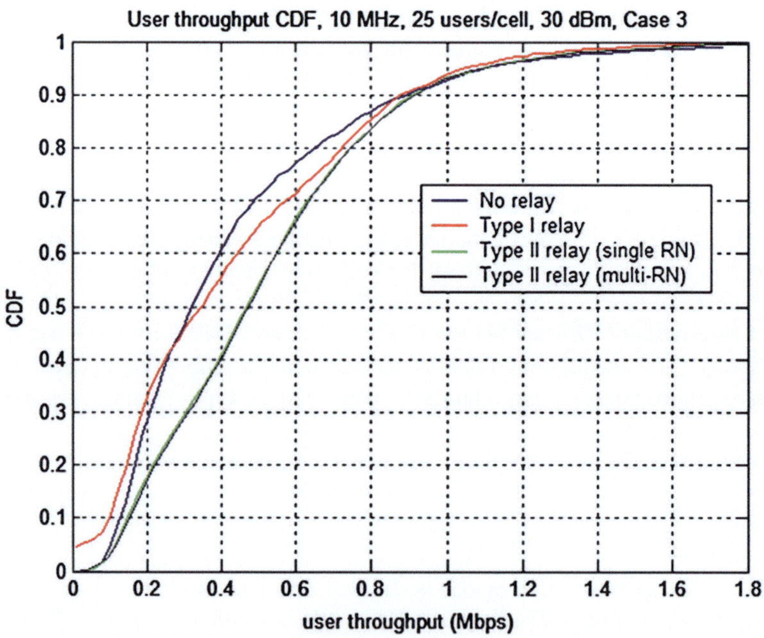

Fig. 3.42 Comparison of user throughput CDFs of different relays, case 3

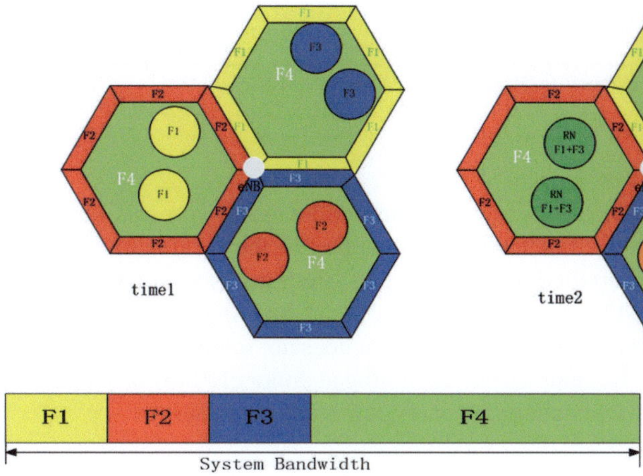

Fig. 3.43 Illustration of the dynamic resource allocation scheme for type 2 relay

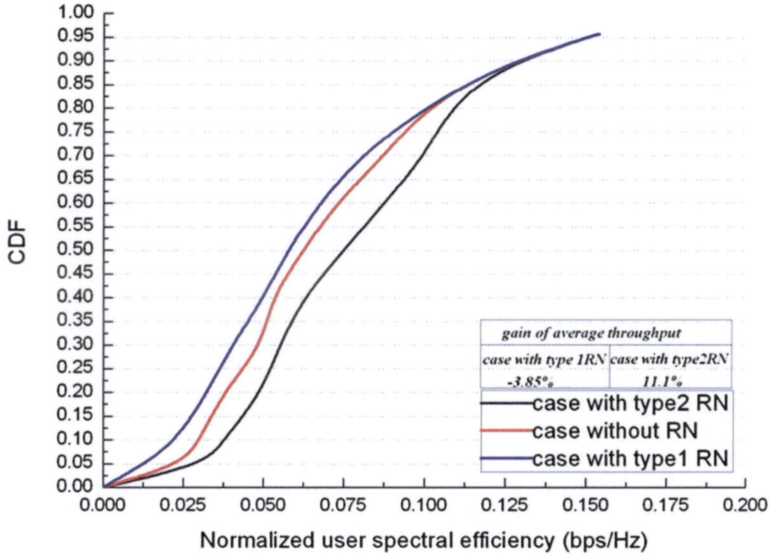

Fig. 3.44 CDF of per user throughput for SFR ICIC method

illustrated in Fig. 3.43 when frequency resources of type 2 relay node are reused in different patterns from time 1 to time 2, Considering the relatively small coverage of each relay node, two subbands used by the cell-edge UEs in other two sectors are reused by the relay node in the sector being considered. With total 50 PRBs for downlink traffic channel, the ratio between F1, F2, F3 and F4 is 8:8:8:26.

Subframe 2# and 3# are reserved for the backhaul transmission. There are four RNs in each sector (cell) even placed on a ring of radius 0.24 ISD.

Table 3.8 System simulation results with SFR ICIC method

	No relay	Type 1 relay	Type 2 relay
Cell average spectral efficiency (b/Hz/s)	1.74	1.67 (−3.9 %)	1.93 (11 %)
Cell edge spectral efficiency (b/Hz/s)	0.0254	0.0142 (−44 %)	0.0327 (28 %)

Figure 3.44 and Table 3.8 show the throughput results for this scenario under 3GPP Case 1. Here the PF scheduling is applied in the simulation. Besides, SFR based ICIC method is applied to improve the system performance for Type 2 relay. For Type 1 relay, no SFR based ICIC method since the implementation will increase the burden of backhaul link transmission. That does not seem to be an issue in Type 2 relay which relies on the centralized scheduling at donor eNB to coordinate eNB and relay transmission. Significant gains of Type 2 relay are observed.

3.6 Other Related Technologies in LTE-Advanced

During the standardization of LTE-A relay backhaul link, downlink reference signals are one important specification area. Some schemes of downlink reference signal (RS) for macro eNB to UE connection can be reused for relay. However, relay backhaul link has its own characteristics, for example, punctured OFDM symbols due to relay timing, R-PDCCH structure and resource mapping, etc.

As a category of low power nodes in heterogeneous networks (HetNet), the introduction of relay node makes interference scenarios more complicated, just as pico-cell or home eNB. Interference coordination schemes of Release 10 enhanced ICIC and some proposals in Release 11 CoMP are generally applicable to relay.

3.6.1 Downlink Reference Signals

Figure 3.45 shows Release 10 downlink demodulation reference signal (DMRS) and CSI-RS with normal cyclic prefix (CP). Release 8 UE specific RS is not configured in this figure. For DMRS, 12 REs are used per PRB pair with orthogonal code cover (OCC) length = 2 when rank = {1, 2} per user. Such case also includes MU-MIMO operation with up to 4 layers. When rank = {3, 4}, 24 REs are used per PRB pair with OCC = 2. When rank = {5, 6, 7, 8}, 24 REs are used per PRB pair with OCC = 4 of two-dimensional code orthogonality. PRB bundling is supported to improve the channel estimation with DMRS.

CSI-RS is for CSI measurement in DL MIMO and CoMP. Only Release 10 CSI-RS can be used for CSI feedback in the new DL transmission mode of Release 10 MIMO. CSI-RS is cell-specific and unprecoded. Its periodicity is 5 ms or multiple of 5 ms. The channel observed from CSI-RS ports can be completely different from the one observed from CRS ports, due to the possible antenna

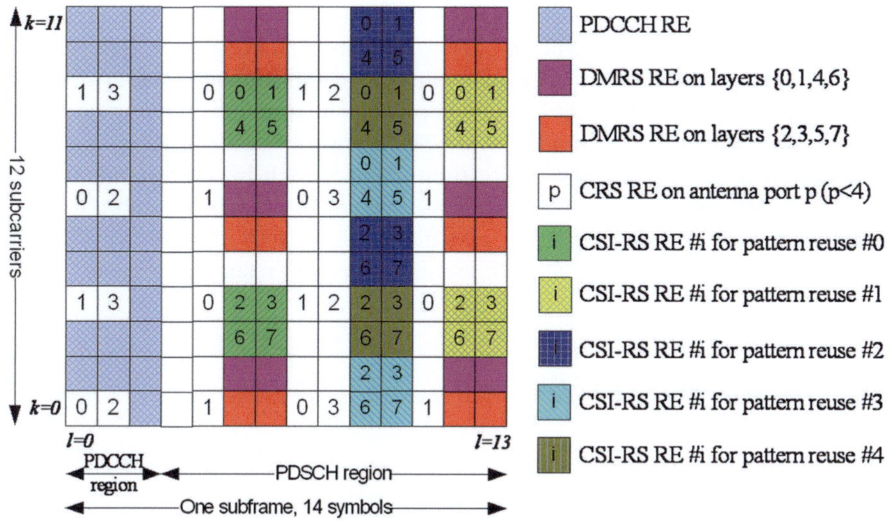

Fig. 3.45 Release 10 LTE downlink reference signal patterns without release 8 UE specific reference signal

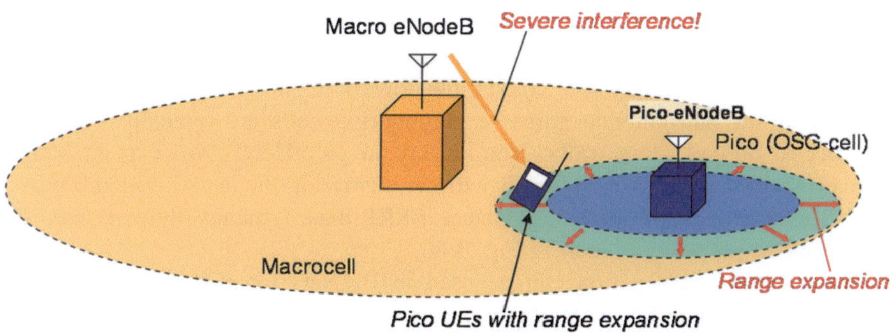

Fig. 3.46 Illustration of range expansion HetNets bias towards pico

virtualization implemented on CRS. The number of CSI-RS ports can be 1, 2, 4, or 8. The density is 1 RE per PRB per port. Every two CSI-RS REs {2i, 2i + 1} shown in Fig. 3.45 are CDM in time-domain. Different CSI-RS pattern reuse in Fig. 3.45 can be assigned to different cells.

When Release 8 UE specific RS is configured, CSI-RS patterns are slightly different from Fig. 3.45, while Release 10 DMRS pattern in Fig. 3.45 keeps unchanged. Following the numbering convention in LTE/LTE-A for RS ports, ports 7–8 are designated to both Release 9 DMRS and Release 10 DMRS. Ports 9–14 are for Release 10 DMRS. Ports 15–22 are for Release 10 CSI-RS.

In Release 10, the CSI-RS is cell specific and would be used for CSI measurement of UEs of a cell. However, the CSI-RS transmission would cause interference to UEs in neighboring cells as the CSI-RS would occupy the resource elements of PDSCH of neighboring cell UEs communicating with their serving cells. This issue becomes more serious when joint transmission is considered in CoMP. Therefore, CSI-RS muting was proposed. The muting configuration is cell specific and signaled via higher layer. The intra-subframe location of muted resource elements is indicated by a 16-bit bitmap, each bit corresponding to a 4-port CSI-RS configuration. All resource elements used in a 4-port CSI-RS configuration set to 1 are muted, i.e., zero power assumed at UE. A Release 10 UE would assume PDSCH rate matching around the muted resource elements.

3.6.2 Enhanced ICIC

The basic idea of interference coordination is to partition and coordinate resource (time, frequency, power, etc.) among cells. Release 8/9 ICIC focuses on the traffic channel and relies primarily on fractional frequency reuse or transmit power management to improve the cell edge performance. Release 10 eICIC provides the mechanisms to make control channel more robust.

Another motivation of Release 10 eICIC is to tackle more severe interference when low power nodes are added to the traditional homogeneous networks, especially if range expansion is performed. In Release 8/9, cell selection is based on RSRP, and effectively the link of the lowest pathloss is selected since the transmit power and antenna gains of neighboring cells are typically the same. In order to offload more traffic from macro to pico/HeNB, we can artificially increase the coverage of pico/HeNB without increasing the actual transmit power level. Range expansion encourages large RSRP bias, typically towards pico for traffic offload purpose, as seen in Fig. 3.46.

While carrier aggregation can be used in HetNets for ICIC purpose, eICIC specification assumes co-channel deployment where the coordination is achieved in time domain by almost blank subframes (ABS) configuration. Almost blank subframe is a new type of subframe that contains no data except essential signals for legacy support, such as synchronization signals (PSS/SSS), primary broadcast channel (PBCH) and common reference signal (CRS). Because of these, a node would cause very low interference to UEs of neighboring cells/nodes when the subframe is configured ABS. The configuration is through a bitmap of 40 bits, thus the period of ABS pattern is 40 ms. The ABS configuration can be semi-static for pico-cells that are connected to the neighboring eNBs or pico-cells via X2 interface. For HeNBs, the ABS configuration is typically static, i.e., via OAM-based solution.

Figure 3.47 shows an example of configurations of ABS in macro and pico cells in FDD. In the macro cell, subframes #1, #5, #9 in the first radio frame and subframes #3, #7 in the second radio subframe are configured ABS (white), meaning that the macro cell DL transmission will impose little interference to UEs in the pico cell. Therefore, the pico cell is encouraged to transmit DL data in those subframes (dark

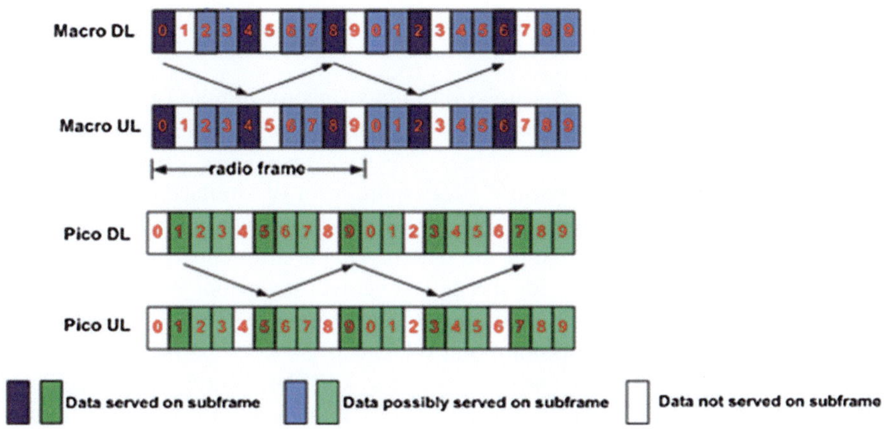

Fig. 3.47 An example of ABS configurations for macro and pico cells in FDD

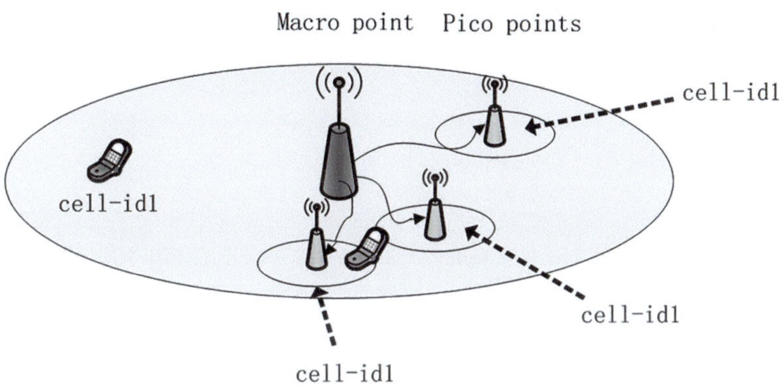

Fig. 3.48 CoMP scenario 4 in release 11 LTE study

green). The ABS pattern is deliberately chosen to fit Release 8 HARQ timing, indicated in the arrows. To avoid causing interference to nearby macro UEs, the pico cell can configure subframes #0, #4 and #8 in the first radio frame and subframes #2 and #6 in the second radio frame to be ABS (white), where the macro cell is encouraged to send DL data (dark blue). In other subframes (light blue or light green), both the marco cell and the pico cell can send DL data, if the transmission would not incur undesirable interference each others' UEs.

While aggressive range expansion can offload more traffic to pico cell or HeNBs, the essential signals such as PSS/SSS, PBCH and CRS would experience very severe interference which cannot be mitigated by using ABS. Then the interference cancellation schemes become necessary at the mobiles. The related work will be studied in Release 11 as the further enhancement for co-channel deployment of HetNets.

3.6.3 CoMP

In CoMP Scenario 4, all the transmission points (macro, remote radio heads and picos) within the coverage of the macro share the same cell id, as illustrated in Fig. 3.48.

In some sense, CoMP Scenario 4 bears quite a lot of similarities with type 2 relay. For example, adding same cell id pico nodes or remote radio heads (RRHs) would not increase the coverage of the macro. This is different from type 1 relay where the primary purpose is for the cell coverage extension. Also, the remote radio head or the same cell ID pico appears transparent to the UEs in the macro cell, at least from DL transmission point of view.

To facilitate PDSCH transmission in Scenario 4, CSI-RS configuration needs to be UE specific, rather than cell specific as in homogeneous deployment. This is also similar to type 2 relay DL where the UE specific DMRS is used in the second hop for either cooperative transmission or resource reuse. Their only difference is that CSI-RS is for CSI feedback, while DMRS is for data demodulation. CRS is not used (not even transmitted from the RN) in the second hop of type 2 relay, nor used for CoMP Scenario 4.

References

1. Europe advanced radio interface technologies for 4G systems (Artist 4G): Advanced relay technical proposals, V2.0, Feb 2011
2. 3GPP R1-082397: Discussion on the various types of relays, RAN1 #53bis, Panasonic, June 2008
3. 3GPP R1-106182: Combination of carrier aggregation and relay in Rel-10, RAN1 #63, NTT DoCoMo, Nov 2010
4. 3GPP R1-083810: Operation of relays in LTE-A, RAN1 #54bis, Qualcomm, Sept 2008
5. 3GPP TR 36.814: Evolved Universal Terrestrial Radio Access (E-UTRA): Further advancements for E-UTRA physical layer aspects
6. 3GPP R1-094449: On the use of PDCCH for relaying, RAN1 #58bis, Ericsson, Oct 2009
7. 3GPP R1-092468: Control channel of backhaul link, RAN1 #57bis, ZTE, June 2009
8. 3GPP R1-094833: Relay performance evaluation with latest channel models, RAN1 #59, Motorola, Nov 2009
9. 3GPP R1-104343: System evaluation of relay performance, RAN1 #62, CATT, Aug 2010
10. 3GPP R1-101919: Type-1 relay performance for uplink, RAN1 #60bis, Nokia, Nokia Siemens Networks, Apr 2010
11. 3GPP R1-100541: Discussion on type 2 relay HARQ issues, RAN1 #59bis, ZTE, Jan 2010
12. 3GPP R1-090775: TDMA H-ARQ Code for Layer-2 Relay in LTE-Advanced, RAN1 #56, Alcatel-Lucent, Feb 2009
13. 3GPP R1-090707: Relaying with channel resource reuse and SIC for LTE-Advanced, RAN1 #56, Fujitsu, Feb 2009
14. 3GPP R1-091709: Performance update of type 1 and type 2 relays, RAN1 #57, ZTE, May 2009
15. 3GPP R1-100442: Performance evaluation for type 2 relay, RAN1 #59bis, Alcatel Shanghai Bell, Alcatel-Lucent, Jan 2010

Chapter 4
Physical Layer Standardization of Release 10 Relay

4.1 Scenario

During relay study item, simulation results of type 1 relay showed some potential gains in cell average capacity and cell edge throughput. As discussed in Chap. 3, there are quite variations in system performances between different companies, and the gains are very sensitive to RN's locations, modeling of backhaul transmissions, interference coordination schemes used.

As the study item drew its close by December 2009, the focus of relay was narrowed down to inband relay for coverage extension scenario. The rationale is as follows:

- The main advantage of relay is the wireless backhaul which allows flexible deployments. In many urban or remote areas, laying down wired backhaul is not realistic or even forbidden due to the strict zone regulations. Yet, these areas often see coverage holes where relay is a strong candidate solution.
- Relay, compared to repeaters, not only provides better performance in terms of SINR improvement, but also reduces the need for Tx and Rx separation. Such separation requirement for repeater limits the deployment choice, especially in crowded cities.
- Relay, compared to microwave backhaul, does not require a separate RF chain backhaul, and is not susceptible to the adverse effect of rain/snow/fog and the block of LOS propagation.
- Inband relay, compared to outband relay, is very attractive to those operators who have limited spectrum for LTE deployment.
- Inband relay proves to be cost effective in improving the coverage in urban areas, according to some CAPEX and OPEX analysis based on simulations using real network setups [1].

Y. Yuan, *LTE-Advanced Relay Technology and Standardization*, Signals and Communication Technology, DOI: 10.1007/978-3-642-29676-5_4,

4.2 Physical Layer Control Channel Specification

Physical layer specification includes the following major aspects:

- Backhaul downlink control channels.
- Relay frame timing.
- Reference signals for backhaul.
- Backhaul subframe configurations and HARQ.

There are some minor modifications in UL control channels and added constraint on DL MIMO transmission in the relay backhaul. The discussion of backhaul control channels began with the necessity of R-PCFICH and R-PHICH, as their specification would affect the basic structure of R-PDCCH.

- R-PCFICH

PCFICH provides the L1 signaling to notify the UEs the number of OFDM symbols for PDCCH in a subframe. PDCCH spans the entire bandwidth and its overhead in frequency domain is fixed. Hence, PCFICH is the only mechanism to adjust the overhead of PDCCH. For the relay control area, the R-PCFICH would be used to indicate the size in time and/or frequency domain (depending on the outcome of the R-PDCCH placement discussion), similar to Release 8 PCFICH. However, R-PDCCH occupies only partial system bandwidth and the overhead of R-PDCCH is relatively stable, because of the limited number of RNs per cell compared to that of UEs and good channel condition of the backhaul with fixed relays.

Alternatively, R-PDCCH resources can be notified by higher-layer signaling. With higher-layer configuration, RN would know the resource area to search for its R-PDCCH. Through blind decoding, RN can find its R-PDCCH within that search area if donor eNB schedule R-PDCCH transmission. Otherwise, the resources can be dynamically used for PDSCH.

With the above consideration, *it was decided not to support R-PCFICH in Release* 10 *relay backhaul.*

- R-PHICH

In Release 8, UL retransmission can be non-adaptive and adaptive. The non-adaptive retransmission relies on ACK/NACK in PHICH, without the triggering of a retransmission grant. The adaptive retransmission is facilitated by retransmission grant in PDCCH, when the UE ignores DL ACK/NACK delivered in PHICH upon its detection of a retransmission grant. NDI bit in a retransmission grant can be interpreted as an implicit DL ACK/NACK as UE retransmits the previously transmitted codeword if NDI bit is not toggled.

For relay backhaul, less fluctuation in channel quality is expected due to the fixed deployment of RN. The stable transmission environment as well as RN site optimization, directional antenna in the backhaul link, low penetration loss, LOS dominant propagation, reduces the HARQ retransmission probability of PUSCH in

Table 4.1 Overhead comparison for UL retransmissions in the backhaul

HARQ retransmission probability (p)	Overhead when R-PHICH exists (N1)	Overhead when R-PHICH does not exist (N2*p)
0.01	3	0.36
0.03	3	1.08
0.05	3	1.80
0.07	3	2.52
0.09	3	3.24

general. Using NDI field in UL grant alone to signal the ACK/NACK may be sufficient. Depending on the operating point of backhaul uplink, i.e., BLER after first transmission, the overall overhead of using UL grant for ACK/NACK would be comparable to that of using R-PHICH. Another consideration is that with limited the number of RNs per cell, there would be less chance of bundled ACK/NACK for efficient R-PHICH transmission.

UL backhaul retransmission probability is a key to determine the necessity of R-PHICH. In [2], the signaling overhead of R-PHICH is analyzed. "p" denotes the retransmission probability, and N1 denotes the number of resource elements used for R-PHICH for each RN, and N2 denotes the number of resource elements used for R-PDCCH for each RN.

The overhead is counted as N1 resource elements for each RN when R-PHICH exists. When R-PHICH does not exist, the overhead is the amount of the resources occupied by retransmission grants which is given by N2*p under the assumption that decoding would be successful after one retransmission. The overhead is compared in Table 4.1. N1 is set to 12/4, assuming the maximum level of PHICH multiplexing (8 DL ACK/NACK per 12 REs) and two-codeword transmission for each RN. N2 is set to 36, assuming that R-PDCCH size is equal to that of PDCCH with aggregation level of 1. It is observed when the retransmission probability is below 9 %, R-PHICH would not save the control overhead compared to using UL grant to trigger the adaptive retransmission.

With the above considerations, *it was decided not to support R-PHICH in Release 10 relay*, which is captured in [3] as

> *The relay node shall not expect HARQ feedback on PHICH, ACK shall be delivered to higher layers for each transport block transmitted on PUSCH.*

• DCI format 3/3A in backhaul

DCI format 3/3A contains transmit power control bits that can be used for fast power control of uplink traffic channels and control channels. For fixed relay deployment, the backhaul channel is supposed to be more stable than eNB-UE channel. In this sense, fast power control is less beneficial and the slow power control may be sufficient in most deployment scenarios. Therefore, *RAN1 decided not to support DCI format 3/3A.*

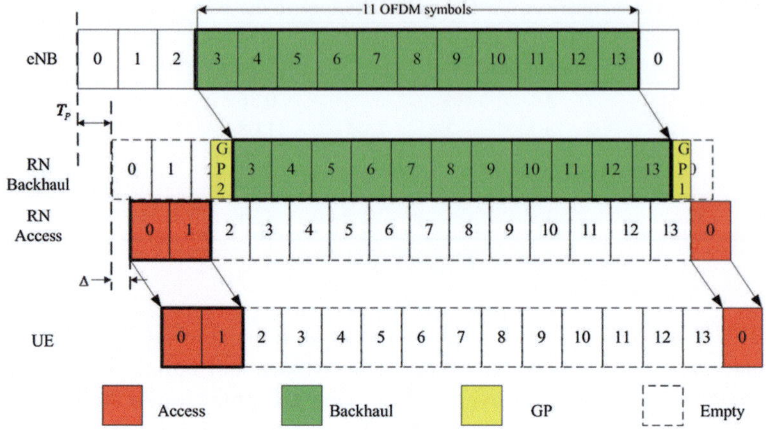

Fig. 4.1 Backhaul subframe structure in relay downlink timing case 1

4.2.1 Relay Downlink Frame Timing

Relay timing is one of first things to decide for relay specification as it determines the backhaul subframe structure for both R-PDCCH and PDSCH. The number of available symbols in a backhaul subframe depends on timing requirement, propagation delay on Un link, RN switching time from transmission to reception or vice versa. According to RAN4 performance specifications, the transmission times are usually in the range of 19–20 μs, which cannot be absorbed by the cyclic prefix typical of 4.7 μs long.

In Release 10 relay specification, two types of timing schemes are supported for backhaul downlink. The first scheme maximizes the symbol utilization in backhaul subframe where access link timing is aligned with backhaul link timing with a fixed offset to account for RN switching time, as seen in Fig. 4.1.

In this example [4], donor eNB's PDCCH has 3 OFDM symbols and RN's PDCCH has 2 OFDM symbols. Due to the propagation delay, there is a "Tp" shift at RN receiver side. In order to receive the last OFDM symbol, RN access link transmission can start "Δ" later than the backhaul link timing from RN receiver prospective, where "Δ" should be at least equal to the RN switching time (GP1). So as long as "Δ + switching time (GP1)" is within an OFDM symbol, only the fourth symbol (#3) is wasted. So in this sample, 11 symbols from #3 to #13 can be used for RN backhaul transmission. If donor eNB's PDCCH has 2 OFDM symbols and RN's PDCCH has 1 OFDM symbol, RN backhaul transmission can start from the third OFDM symbol (#2) and end by the last OFDM symbol (#13, or #6 in the second slot).

The second scheme ensures global synchronization between DeNB and RNs, which is especially important for TDD systems. There, 3 μs synchronization requirement is imposed on neighboring macro sites. Figure 4.2 illustrates an example where donor eNB has 3 OFDM symbols for PDCCH and RN has 2

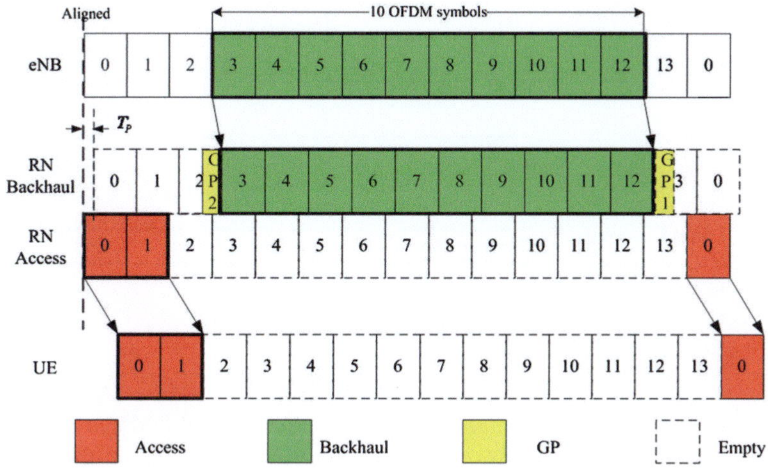

Fig. 4.2 Backhaul subframe structure in relay downlink timing case 3

OFDM symbols for PDCCH. Due to the propagation delay (Tp) and RN switching time (GP1), the last OFDM symbol (#13, or #6 in the second slot) cannot be received by RN. The third symbol (#2) is used for RN switching (GP2) and RN receiving can start from the fourth symbol (#3). If PDCCH of both donor eNB and RN have 1 OFDM symbol and the propagation delay (Tp) is larger than the RN switching time (GP2), RN reception can start from the second symbol (#1).

Counting the different numbers OFDM symbols for donor eNB and RN, and different types of DL relay timing schemes, the backhaul RN reception may start from OFDM symbol #1, #2 or #3, and end by OFDM symbol #13 (#6 in the second slot), or #12 (#5 in the second slot). Note that the case of starting from #1 and end by #13 is not supported, as downlink timing case 1 and timing case 3 cannot coexist in an RN.

In [3], the downlink backhaul subframe structure is specified in the two tables, one for the first slot and the other for the second slot. The table assumes normal cyclic prefix. Such per-slot definition of subframe structure is motivated by TDM + FDM multiplexing of R-PDCCH and PDSCH, which will be discussed shortly after.

The wording "the simultaneous operation of configuration 0 in" Table 4.2 "and configuration 0 in" Table 4.3 "is not supported" prohibits the erroneous configuration.

Not that the end symbol of downlink backhaul subframe is often implementation specific, i.e., whether to operate in FDD or TDD. Therefore, no higher layer signal is required for its configuration.

Also note that for narrow band operation such as 1.4 MHz, PDCCH region can span up to 4 OFDM symbols. So in principle, the start symbol could be the 5th OFDM symbol (index #4). However, in order to simplify the specification, and also considering the small chance of operating 1.4 MHz in LTE-A, it was decided

Table 4.2 OFDM symbols for eNB-to-RN transmission in the first slot

Configuration	DL-startsymbol	End symbol index
0	1	6
1	2	6
2	3	6

Table 4.3 OFDM symbols for eNB-to-RN transmission in the second slot

Configuration	Start symbol index	End symbol index
0	0	6
1	0	5

not to tailor the standards to handle the specific case of 4-OFDM symbol PDCCH. Rather, the situation is to be solved by implementation, i.e., donor eNB scheduler should restrict its PDCCH region within the first three OFDM symbols in backhaul subframes when the network is operating with 1.4 MHz system bandwidth.

4.2.2 Configuration of Start Symbol of R-PDCCH and PDSCH

A topic related to RN downlink timing is how to configure the start symbol for R-PDCCH. Since R-PDCCH carries potential DL and UL grants, including RRC signaling for the backhaul. While higher layer signaling and blind decoding can be used to determine the start symbol of R-PDCCH, the associated complexity in specification, implementation and testing are highly undesirable. To simplify the dimensioning and specification of R-PDCCH, the start symbol for R-PDCCH is fixed to be OFDM symbol #3. Note that the loss of efficiency is very minimal, even assuming the worst usage of backhaul time resources.

The potential vacant OFDM symbols before OFDM #3 in that physical resource block (PRB) should not be used for PDSCH transmission. Otherwise, the resource mapping for PDSCH would be too complicated and fragmented.

In [3], the start symbol of R-PDCCH is specified as

An R-PDCCH is transmitted according to configuration 2 of Table 4.2.

Regarding the start symbol of backhaul PDSCH, there were some discussions on whether to define new L1 signaling in R-PDCCH, or higher-layer signaling for the configuration. Through R-PDCCH L1 signaling, the donor eNB can dynamically signal where PDSCH is started. With higher-layer signaling, the start symbol of PDSCH would be configured or updated in a semi-static fashion.

As discussed in Sect. 4.2.1, the start symbol of the PDSCH depends on the length of PDCCH (in number of OFDM symbols) at the donor eNB, the length of PDCCH at the RN, the propagation delay between eNB and RN. Since the relay location is fixed, the backhaul propagation delay would not change after the relay is deployed.

The length of PDCCH of RN cell does not seem to vary much since there would no PDSCH transmitted in that subframe by RN, and no DL grants either.

The length of PDCCH of donor eNB may vary from subframe to subframe, depending on the traffic demand from RN transmit buffer which is the aggregate of the traffic of multiple UEs served by each RN. It is arguable whether the aggregate traffic would be more constant.

The length of PDCCH of donor eNB is also affected by whether macro UEs are scheduled in the backhaul subframes. Clearly, with more macro UEs co-scheduled, the length of PDCCH of donor eNB would vary dynamically. Therefore, potentially, the start symbol of PDSCH would change subframe by subframe.

The discussion dragged for some time since both approaches have their strong arguments. In the end, considering the specification complexity (to define a new L1 signaling in R-PDCCH) would outweigh the performance benefit (in terms of the saved OFDM symbols for backhaul PDSCH), the higher-layer configuration is adopted. In [3], it is specified as:

> *The parameter DL-StartSymbol in* Table 4.2 *is configured by higher layers.*

4.2.3 Relay Uplink Frame Timing

There were also extensive discussions on UL relay timing. Similar to relay downlink timing, uplink frame timing can target for either maximizing the usage of SC-FDMA symbols in an uplink backhaul subframe, or meeting the exact synchronization requirement between the eNB and the RN. The following schemes were proposed with the above two purposes.

- Uplink timing Case 2b

In this case, all 14 SC-FDMA symbols can be used in backhaul even though the RN switching time is longer than the cyclic prefix as shown in Fig. 4.3; [5]. The access subframe which is in front of a backhaul subframe should be configured as a cell-specific SRS subframe and the last SC-FDMA symbol (#13) in this subframe is punctured for RN switching without impacting on UE traffic transmission.

As shown in Fig. 4.3, RN UL access reception time is advanced with respect to eNB UL backhaul reception time by $T_P - \Delta$ if $T_P \geq \Delta$ ($T_P - \Delta$ denotes a delay, if $T_P < \Delta$), and the fixed delay should satisfy $T_{GP1} \leq \Delta \leq T_{symbol} - T_{GP2}$ for a maximum usage of backhaul resource.

Uplink timing case 2b maximizes the resource utilization in the backhaul subframes. In addition, there is no impact on backhaul link SRS and PUCCH transmission. But it has the impact on access link where "fake" SRS needs to be configured in the preceding subframes to allow appropriate handling of shortened subframes. On the other hand, "true" SRS may still be needed in other subframes in order to support UL frequency selective scheduling. Therefore, SRS configuration should be done carefully to make sure that the impact on the access UL measurement would be small.

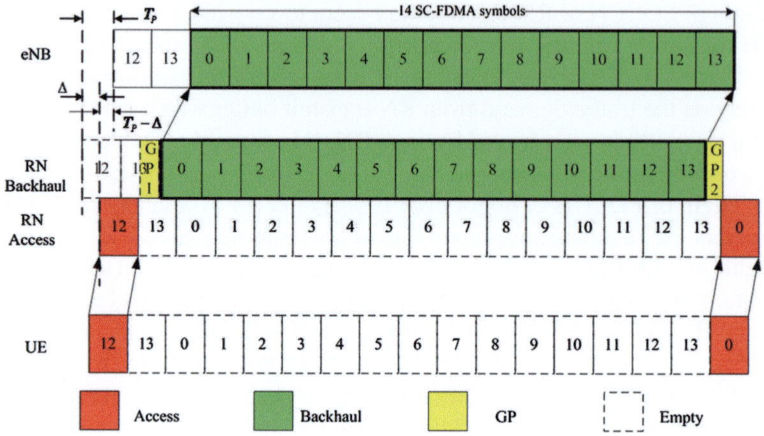

Fig. 4.3 Subframe structure in relay uplink timing case 2b

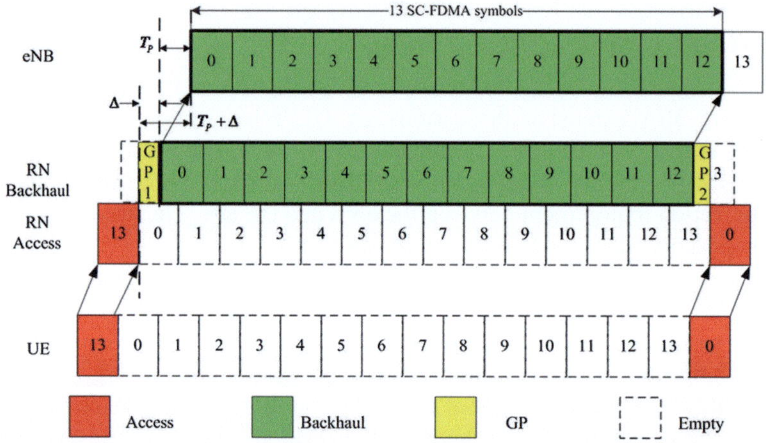

Fig. 4.4 Backhaul subframe structure in uplink timing case 3

- Case 3

As Fig. 4.4 shows, RN UL reception time is advanced with UL transmit time by $\Delta$and as a result RN UL reception time is advanced with eNB UL reception time by $T_P + \Delta$. Similar to case 1, if the fixed gap Δ satisfies $T_{GP1} \leq \Delta \leq T_{symbol} - T_{GP2}$, up to 13 up SC-FDMA symbols could be used for backhaul transmission.

In uplink timing case 3, the last backhaul symbol (#13) is punctured and cannot be used by RN for "true" Un SRS transmission, only the shortened PUCCH is supported by "fake" Un SRS configuration.

- Modified uplink timing case 4

Uplink timing case 4 is mainly used for TDD system where the eNB and RN timing should be strictly synchronized, both for the downlink and the uplink. The original uplink timing case 4 requires puncturing the first SC-FDMA symbol in the UL backhaul subframe, which is not desirable since it involves major changes in Release 8 PUCCH specification. The modified uplink timing case 4 removes that necessity, so that Release 8 PUCCH can be reused.

Similar to uplink timing case 2b, the consequence of saving the first SC-FDMA symbol in the backhaul subframe leads to the puncturing of the last SC-FDMA symbol in the access subframe right before the backhaul subframe. The puncturing can have no specification impact as RN may broadcast cell-specific SRS configuration to prevent PUSCH/PUCCH transmissions on the last SC-FDMA symbol by RN served UEs.

Depending on the eNB-RN distance, modified uplink timing case 4 would lead to three timing relationships [6], two of them shown in Figs. 4.5 and 4.6. In each relationship, there can be two situations: (1) UL backhaul subframe followed by an UL access subframe or (2) UL backhaul subframe followed by a DL access subframe. Figure 4.5 shows the subframe relationship when the eNB-RN propagation delay is less than the typical RN Tx/Rx switching time which is about 20 μs, translated into 6 km. In Fig. 4.5a, the last SC-FDMA symbol has to be punctured to accommodate the Tx/Rx switching time, which is similar to uplink timing case 3. In Fig. 4.5b, all 14 SC-FDMA symbols can be used for the uplink backhaul subframe, since there is no Tx/Rx gap time needed.

Figure 4.6 shows the subframe relationship when the eNB-RN propagation delay is larger than the typical RN Tx/Rx switching time which is about 20 μs, translated into 6 km, and the sum of eNB-RN propagation delay and RN Tx/Rx switching time is less than a SC-FDMA symbol duration, translated into 15 km. Regardless whether the UL backhaul subframe is followed by an UL access subframe or a DL access subframe, all 14 SC-FDMA symbols can be used for the uplink backhaul subframe.

When the eNB-RN is larger than 15 km, the sum of eNB-RN propagation delay and RN Tx/Rx switching time would exceed one SC-FDMA symbol, leading to further loss of SC-FDMA symbols either at the backhaul link or the access link. In such case, the strict requirement for time alignment between eNB and RN may be relaxed to avoid excessive resource waste in the backhaul.

In the light of minimizing the specification impact, *the discussion of UL relay timing was concluded with no specification in RAN*1.

Rather, it is left to the network implementations to ensure whether all 14 SC-FDMA symbols or the first 13 SC-FDMA symbols would be available in the backhaul.

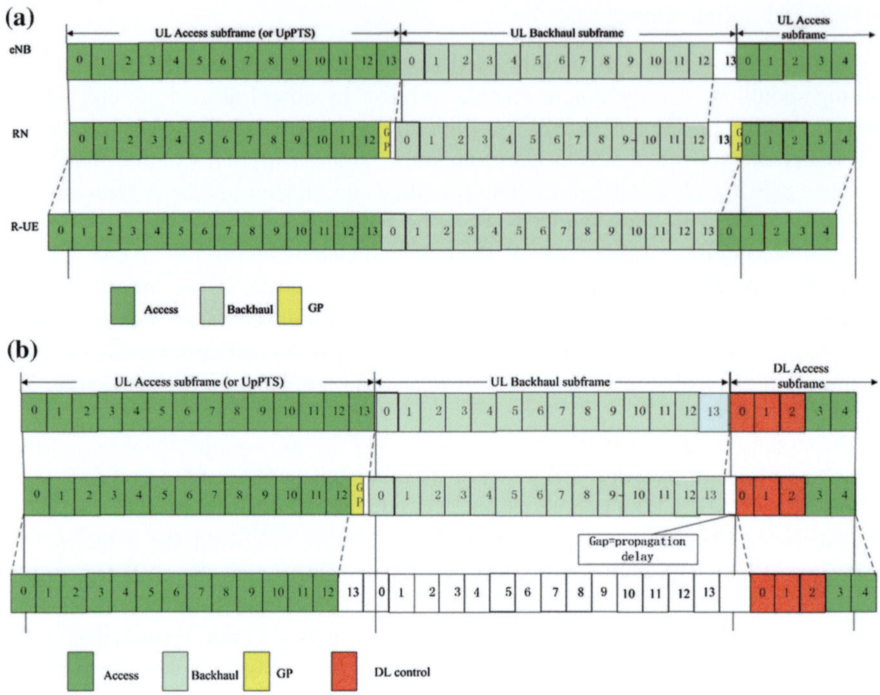

Fig. 4.5 **a** Example for subframe relationship when eNB-RN distance is less than 6 km and UL backhaul subframe is followed by an UL access subframe. **b** Example of subframe relationship when eNB-RN distance is less than 6 km and UL backhaul subframe is followed by a DL access subframe

4.2.4 Relay Node Synchronization

Transmitter and receiver's timing deviation depends on the clock precision on both sides. Receiver will lose synchronization when the clock drifting is beyond certain level. RN needs to adjust its timing and frequency periodically for backhaul transmission.

If backhaul subframes and access link subframes have the same index, Release 8 primary synchronization and sequence secondary synchronization sequence (PSS/SSS) from eNB cannot be directly used for RN. The reason is that FDD subframes {#0, #4, #5, #9} and TDD subframes {#0, #1, #5, #6} have to be configured in relay node to maintain synchronization, paging etc. basic functionalities for relay served UEs. Therefore, donor eNB has to configure other subframes such as {#1, #2, #3, #6, #7, #8} in FDD and {#2, #3, #4, #7, #8, #9} for backhaul transmission which does not have PSS/SSS. There are several potential solutions [7]:

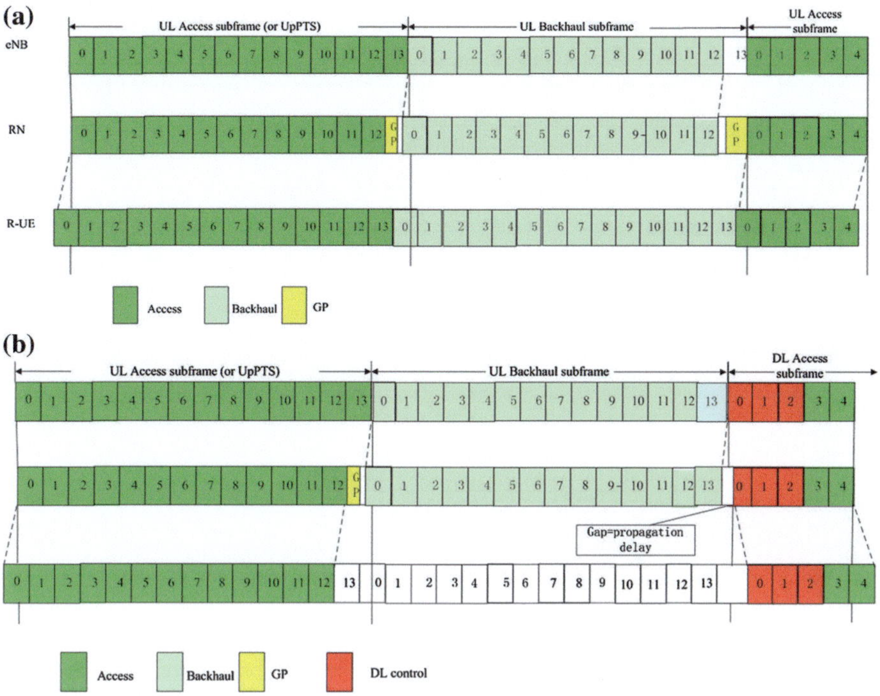

Fig. 4.6 **a** Example for subframe relationship when eNB-RN distance is between 6 and 15 km, and UL backhaul subframe is followed by an UL access subframe. **b** Example for subframe relationship when eNB-RN distance is between 6 and 15 km, and UL backhaul subframe is followed by a DL access subframe

- Scheme 1: using CRS, CSI-RS or GPS

Scheme 1 does not need specification change. However, the sparse distribution of CRS and CSI-RS in time and frequency may not satisfy the stringent requirement of synchronization, assuming that RN's clock should be more precise than UE's clock. The precision limitation would be more pronounced for small system bandwidth operation or for non-stationary RN. It is known that RN may not get the GPS signal due to diversified deployment scenarios (e.g. below the building rooftop or indoor scenario), and GPS scheme also increases the system CAPEX.

- Scheme 2: offset of subframe indexing

It may be possible that the sequence number of subframe on direct link is different from access link for FDD. So configuring a fixed "shift" of access subframe sequence number would avoid the conflict between access subframe #0, 4, 5, 9 and direct subframe #0 and #5 that where PSS/SSS resides. Hence, RN could receive PSS/SSS multiplexing on direct subframe #0 and #5 if at the same the time access subframe is configured as MBSFN. As an example shown in Fig. 4.7, access subframe #1 and #6 can be configured as MBSFN subframes.

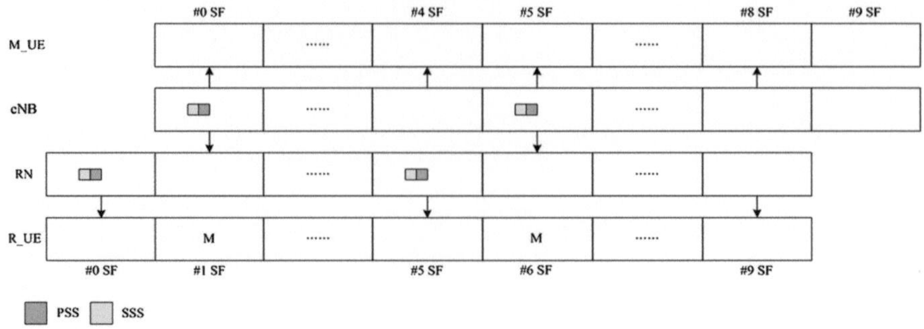

Fig. 4.7 Subframe sequence number shifting

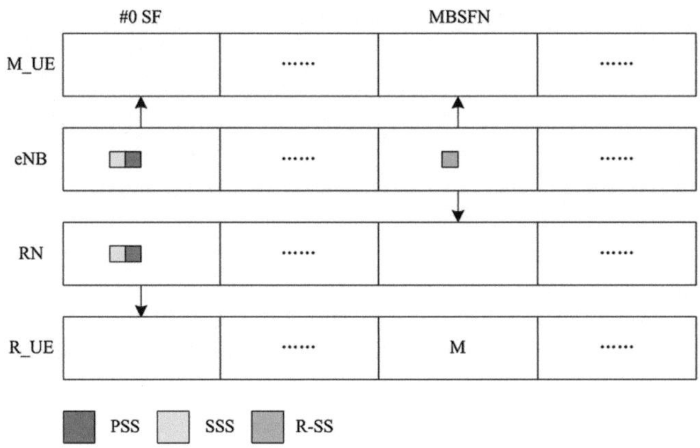

Fig. 4.8 Relay specific synchronization channel

In addition, PBCH from eNB could also be received in direct-link subframe #0 where macro UEs are co-scheduled together with RNs.

While Scheme 2 does not require specification changes, it does limit the donor eNB's scheduling flexibility for backhaul link resource, since certain subframes should be configured as MBSFN. For example, macro UEs and RNs may not be co-scheduled in some subframes.

Also, Scheme 2 is not applicable to TDD systems where the subframe indexing of eNBs and RNs in a network should be the same to avoid the severe inter-node interference.

- Scheme 3: introducing new synchronization channel

Figure 4.8 shows an example of relay specific synchronization channel which is mapped to the last OFDM symbol of the first slot in the backhaul subframe. Since RN has already distinguished FDD or TDD during initial synchronization, the

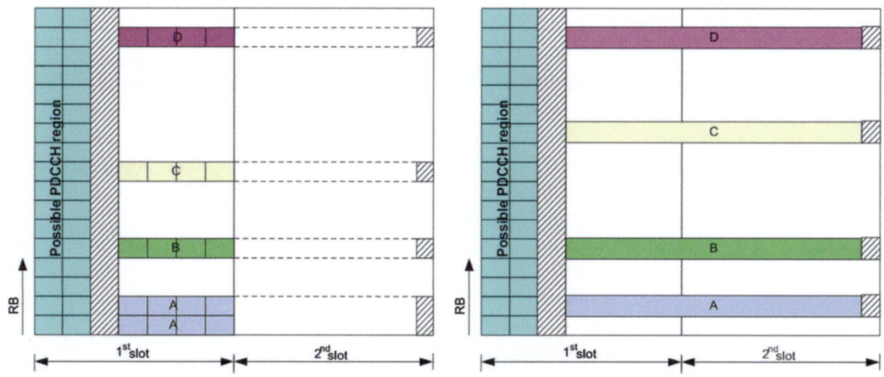

Fig. 4.9 R-PDCCH and PDSCH multiplexing

index of OFDM symbol occupied by relay synchronization channel can be the same for FDD and TDD.

While Scheme 3 has the merit of keeping the flexibility of backhaul scheduling and can be readily used in TDD, it requires new specification of synchronization channel specific for relay backhaul. During that time, relay node synchronization was regarded by many companies to be an implementation issue. Therefore, *no RAN*1 *specification is to address relay node synchronization in Release* 10.

4.2.5 R-PDCCH Multiplexing

The most important aspect of R-PDCCH specification is how R-PDCCH is mapped to PDSCH region, or equivalently, how R-PDCCH and PDSCH are multiplexed. The whole discussions lasted for 5 months in RAN1, from January 2010 till May 2010. All of them centered around two competing multiplexing schemes as illustrated in Fig. 4.9: time division multiplexing + frequency division multiplexing (TDM + FDM) on the left, and pure FDM on the right. Different colors (or "A", "B", "C", "D") denote different relay nodes. The shady regions are for Tx/Rx and Rx/Tx switching.

The pros and cons of TDM+FDM versus pure FDM multiplexing are summarized in Table 4.4.

- Hybrid FDM + TDM

R-PDCCH is transmitted on a subset of physical resource blocks (PRB), or more specifically, only the symbols in the first slot. The remaining resources in that PRB pair can be used to carry PDSCH for relay nodes, but not for PDSCH of macro UEs.

Hybrid TDM + FDM multiplexing has the following advantages: RN energy saving and shorter R-PDCCH decoding latency. In principle, RN can rely on CRS

Table 4.4 Comparison of TDM + FDM versus pure FDM multiplexing of R-PDCCH

	Hybrid TDM + FDM	Pure FDM
Decoding latency	Short, less requirement for baseband processing power in RN	Long, significant burden on the baseband processing time budget
Resource granularity	Finer, e.g., one PRB for one RN's R-PDCCH, suitable for cross-interleaved R-PDCCH	Need to be paired between RNs to fill up a PRB
Frequency diversity and selectivity	Suitable for frequency diversity	Suitable for frequency selectivity
R-PDCCH/ PDSCH multiplexing	Less flexible, resulting in more complicated PDSCH resource mapping	More flexible, less issue with PDSCH

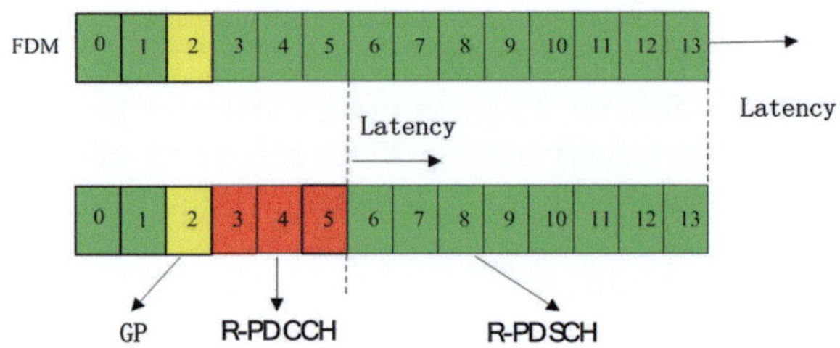

Fig. 4.10 A comparison of R-PDCCH decoding latency

or DM RS in the first slot, thereby saving energy by turning off the receiving chain. It can be seen from the example in Fig. 4.10 that R-PDCCH decoding latency is about eight OFDM symbol shorter compared with pure FDM, if CRS up to OFDM symbol #4 are used for R-PDCCH detection.

It should be pointed out that the exact calculation of decoding latency depends on the receiver implementation at RN. When Release 10 DMRS is used for R-PDCCH, one can either use DMRS in the first slot, or in both slots to estimate the backhaul channel. Using DMRS only in the first slot would allow early detection of R-PDCCH in the first slot. However, using DMRS in both slots would improve the channel estimation accuracy, albeit losing the benefit of early detection of R-PDCCH.

TDM + FDM method tends to provide more frequency and interference diversity levels, compared to pure FDM method. It is noted that as the number of PRBs for R-PDCCH increases, frequency diversity can also be achieved in pure FDM method.

TDM + FDM would complicate the resource scheduling in the backhaul subframe as the resources are essentially in two dimensions, slot number and PRB index. Also, DM-RS patterns and ranks have to be carefully designed and specified to ensure compatibility of R-PDCCH and PDSCH when they coexist in one PRB pair.

- Pure FDM

Pure FDM allocates a certain number of PRBs exclusively for R-PDCCH transmission. Therefore, R-PDCCH is cleanly separated from PDSCH, regardless whether the PDSCH is for macro-UEs or for RNs. R-PDCCH and PDSCH multiplexing becomes more straightforward, and resource scheduling can be less complex. Similarly, the same DM-RS patterns can be used for PDSCH for R-PDCCH demodulation.

Power control of Release 8 PDCCH helps to achieve coverage when CCE aggregation level cannot be set too high, either because the maximum aggregation level is reached, or it would cause severe blocking. Power control may also be necessary for R-PDCCH when an RN site is not well planned. In FDM method, there is more flexibility for power control, compared to FDM + TDM case where different downlink transmit powers would be needed for OFDM symbols for R-PDCCH and R-PDSCH of a given UE. If DM-RS is used, it would be very difficult to transmit R-PDCCH and PDSCH with different powers.

The pure FDM method requires more memory at RN receiver, since more data need to be buffered before detecting DL grant in R-PDCCH.

It is not straightforward to argue the specification complexity between TDM + FDM and pure FDM approaches. TDM + FDM approach is heavily based on Release 8 PDCCH design philosophy. Therefore, most of PDCCH specification could be borrowed. However, the mapping PDSCH only to the second slot is something new and definitely requires certain specification work. Pure FDM approach is quite different from TDM based PDCCH and hence needs new specification. But a closer look at would reveal that the actual work for specification may not be that burdensome as it appears, since Release 8 PDCCH mapping to resource elements can still be reused in FDM approach.

As a compromise, RAN1 decided to support both TDM + FDM and pure FDM schemes in Release 10 relay. To reduce the decoding latency in backhaul downlink in the case of FDM multiplexing, DL grants are only allowed to be sent in the first slot of a PRB, and UL grants are only allowed to be sent in the second slot. TDM + FDM can be used when only DL grant is sent in a PRB, so that the second slot can be used for PDSCH of RN(s). Since R-PDCCH starts from the fourth OFDM symbol (#3), there are four OFDM symbols for DL grant(s) per subframe.

In [3], there are several places specifying TDM + FDM and pure FDM for R-PDCCH. R-PDCCH configuration granularity is at slot level as seen in Tables 4.2 and 4.3 for the first slot and the second slot, respectively. The wording "An R-PDCCH is transmitted according to configuration 2 of" Table 4.2 "or configuration 0 or 1 of" Table 4.3 indicates that R-PDCCH can be transmitted in the first slot, or the second slot, or in both slots of a PRB pair.

Since R-PDCCH and PDSCH are complementary in the backhaul resource map, the support of both TDM + FDM and pure FDM for R-PDCCH is also specified from PDSCH resource point of view as the PDSCH shall only be mapped to resource elements in OFDM symbols configured according to Tables 4.2 and 4.3. the PDSCH shall not be mapped to any resource element in the first slot of an RB pair when the first slot of the RB pair is used for R-PDCCH transmission.

A relay node shall upon detection of an R-PDCCH intended for the relay node in a subframe, decode the corresponding PDSCH in the same subframe with following assumptions.

- If the relay node receives a resource allocation which overlaps a PRB pair in which a downlink assignment is detected in the first slot, the relay node shall assume that there is PDSCH transmission for it in the second slot of that PRB pair.
- For a PRB pair where the relay node detects at least part of a downlink assignment in the first slot, the relay node shall assume that the first slot of the PRB pair is not used for PDSCH transmission.

In Release 8 resource allocation type 0, the resource unit is resource block group (RBG). One RBG contains 1, 2, 3 or 4 PRBs for system bandwidth of 1.4 ~ 1.4, 3–5, 10 and 15–20 MHz, respectively. There can be multiple PRB pairs in a RBG where R-PDCCH only occupies the first slots of some PRBs, but not all PRBs in a RBG. Hence, some clarification is needed on how to multiplex R-PDCCH and PDSCH in that situation. There were some discussions on the possible new schemes for resource indication when a RBG contains multiple PRBs [8]. Some proposals recommended redefining the resource allocation type 0 for relay backhaul, in order to fully utilize the residual resources left by the partial occupation of R-PDCCH in a RBG. No conclusion was reached on this issue as many companies believed that maximizing the resource usage under such condition can be done by implementation, i.e., appropriate resource scheduling. Consequently, Release 8 resource allocation type 0 is reused in relay backhaul.

It is still worthwhile to clarify how the resource is mapped in the above situation. Let us consider a RBG with 4 PRB pairs and first look at the case when R-PDCCHs of different RNs are not cross-interleaved. When the resource allocation (RA) bit is 1 for RN1 and the DL grant occupies only one slot of PRB pair as shown in Fig. 4.11a; [9], Un PDSCH would be transmitted on the available resources of total 7 slots, in the assigned RBG. However, for RA = 0, it means no PDSCH in this RBG for RN1. When the entire resources of a RBG contain DL/UL grants of different RNs, as shown in Fig. 4.11b, the multiplexing can be made at finer granularity, i.e., 1 or 2 PRBs instead of 4 PRBs. In the case that resource allocation (RA) bits are all zeros for RN1, RN2 and RN3, those RNs would know that the second slots either have no PDSCH, or carry each RN's UL grant.

When different RNs' R-PDCCHs are cross-interleaved, full utilization of second slot resources can be achieved by proper scheduling at eNB. Figure 4.12a shows an example where DL grants of RN1 and RN2 are cross-interleaved and transmitted in the first slots of all PRBs in a RBG, the second slots of all PRBs in this RBG can be used for PDSCH transmission for RN1 if RA = 1 for RN1 and RA = 0 for RN2. An eNB scheduler can also fill up an entire RBG with DL grants and UL grants for multiple RNs as seen in Fig. 4.12b, when RA = 0 for both RN1

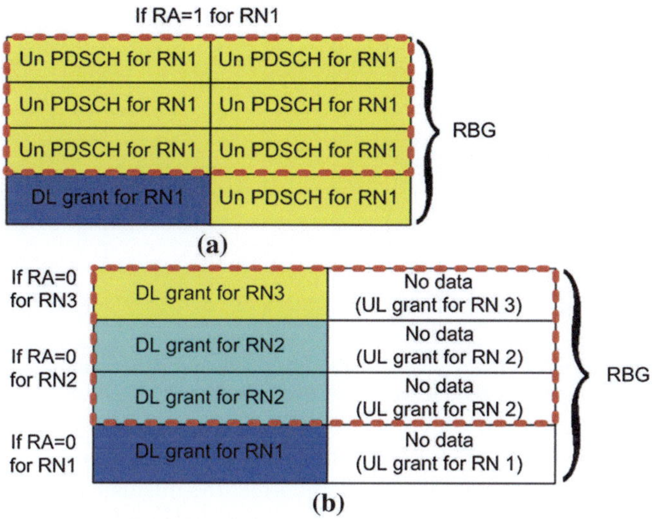

Fig. 4.11 R-PDCCH and PDSCH multiplexing details in one RBG, without cross-interleaving

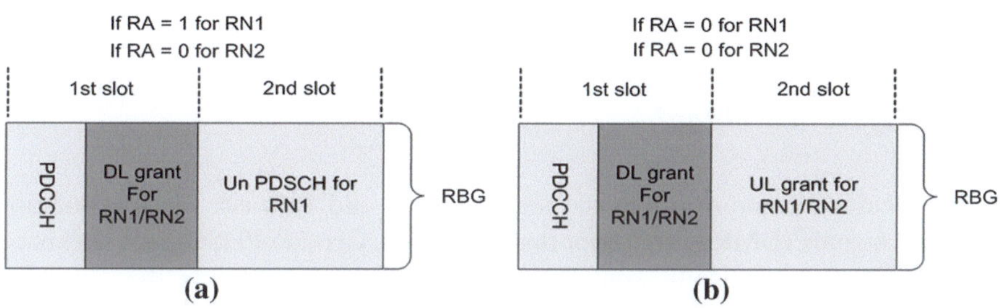

Fig. 4.12 R-PDCCH and PDSCH multiplexing example, with cross-interleaved R-PDCCHs

and RN2. Such configuration is more suitable when: (1) the R-PDCCH resources are more balanced between the first slot and the second slot, i.e., 4 OFDM symbols in the first slot and 6 OFDM symbols in the second slot, and (2) DL grant payload is similar to UL grant payload.

As Fig. 4.11a shows, the number of avaiable resource elements for PDSCH is decreased since R-PDCCH carves out certain slots of PRB pairs from a whole RBG. The transport block size (TBS) defined for eNB-UE transmissions may not accurately fit the resource mapping of PDSCH for backhaul, leading to less efficient link adaptation in the backhaul link. New scaling factors were proposed to get suitable TBS [10]. However, since PDSCH patterns for RN in the backhaul are quite diverse, i.e., there would be numerous combinations of R-PDCCH and PDSCH resource multiplexing, the optimization of new scaling factors requires extensive work. It is also noted that backhaul channel usually does not change fast

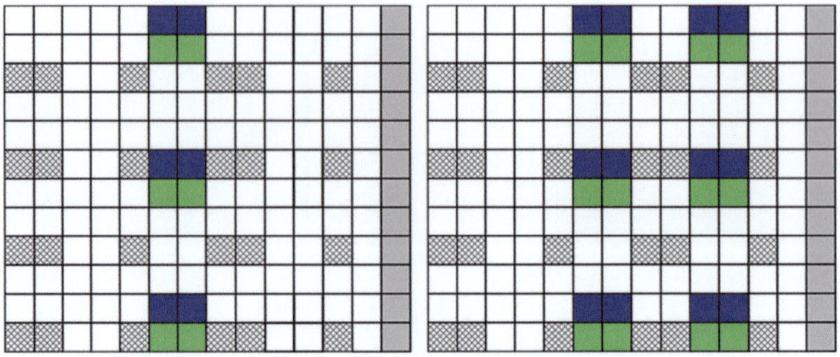

Fig. 4.13 DMRS patterns of two alternatives in relay downlink frame timing case 3, i.e., eNB and RN timing should be strictly aligned, left pattern is adopted in Release 10

due to the fixed deployment of RNs. Outer loop link adaptation, although slow to converge, may be able to mitigate and self correct the mismatch of transport block size. So no new scaling factors were specified, i.e., *Release 8/9/10 transport block size tables are reused in backhaul PDSCH and PUSCH transmission*. It is up to the scheduler to handle those situations. The scheduler implementation aspect will be further discussed in Chap. 6.

4.2.6 Reference Signal

Both Release 8 common reference signal (CRS) and Release 10 demodulation reference signals (DMRS) are supported for R-PDCCH and PDSCH for backhaul. Donor eNB semi-statically configures the reference signal type for backhaul downlink. Note that Release 10 DMRS should solely be used when MBSFN subframes are configured for backhaul.

As discussed in Sect. 4.2.2, downlink relay frame timing case 3 requires strictly aligned timing between eNB and RN, which makes the last OFDM symbol in the backhaul subframe not available to an RN. However, Release 10 DMRS spans both slots and is present in the last OFDM symbol (#6) in the second slot. To solve this issue, two alternatives were proposed as illustrated in Fig. 4.13. Their comparisons are highlighted in Table 4.5.

The reduced DM-RS pattern (Alt.1) can only support up to 4 layer transmission on backhaul link because it only support length-2 orthogonal code cover (OCC) in time domain for each coded division multiplexing (CDM) group (following DM-RS design for normal CP). However, not supporting 5–8 layer transmission on backhaul link is not seen as a major restriction, since it is not expected that relay nodes will have more than 4 antennas. Furthermore, relays will be deployed in optimized spots with high LOS probability, so that only up to 4 layer transmission would be more feasible. For donor eNBs equipped with 8 antennas, MU-MIMO transmissions can still be carried out to serve multiple relay nodes simultaneously.

Table 4.5 Comparison of reduced (Alt. 1) and shifted (Alt. 2) DMRS patterns

	Reduced (agreed)	Shifted (rejected)
Supported rank for PDSCH of RN	Up to 4	Up to 8
Impact on release 10 CSI-RS	No impact	Extra CSI-RS patterns defined for backhaul
Performance	Notable degradation in mobile environment	Good
DMRS overhead	Less	Moderate

Table 4.6 Link level simulation parameter to evaluate performance of PDSCH when reduce DMRS and shifted DMRS are used

Number of antennas	2 × 2, 4 × 4 uncorrelated antenna
Carrier frequency	2.0 GHz
System bandwidth	10 MHz
Scheduled subcarriers	Contiguous 6 PRBs
(R-)PDSCH	10 OFDM symbols
Cell-specific RS	4 ports
Channel model	Pedestrian A
Speed	3 km/h
Rank adaptation	Fixed rank (1 and 2)
Precoding	Closed-loop spatial multiplexing (Rel 8)
Link adaptation	On
Channel estimation	2D-MMSE filter
MIMO detection algorithm	MMSE
Channel coding	Turbo code
HARQ	On

The shifted DM-RS pattern (Alt.2) allows rank 8 transmission in the backhaul. However they would collide with CSI-RS if configured in the backhaul subframe. As seen from Release 10 CSI-RS patterns in Fig. 3.45, the shifted DM-RS in OFDM symbol #9 and #10 would collide with CSI-RS, whose reserved position is also in OFDM symbol #9 and #10. In that case, the shifted DM-RS pattern would have to be punctured so that the donor eNB can transmit CSI-RS in an undisturbed manner.

The shifted DM-RS pattern (Alt.2) is believed to have performance advantage, when the extra overhead is not taken into account. Channel estimation would be more accurate with the shifted DM-RS pattern (Alt.2) because more RS are available. On the other hand, more resource elements would be available for PDSCH transmission with reduced DM-RS pattern (Alt. 1), which helps to boost the backhaul capacity. Some simulation was carried out [11] to study the performance of both alternatives. The simulation parameters are listed in Table 4.6. The link level simulation results are shown in Fig. 4.14a and b for 2 × 2 and 4 × 4 antenna configurations, respectively.

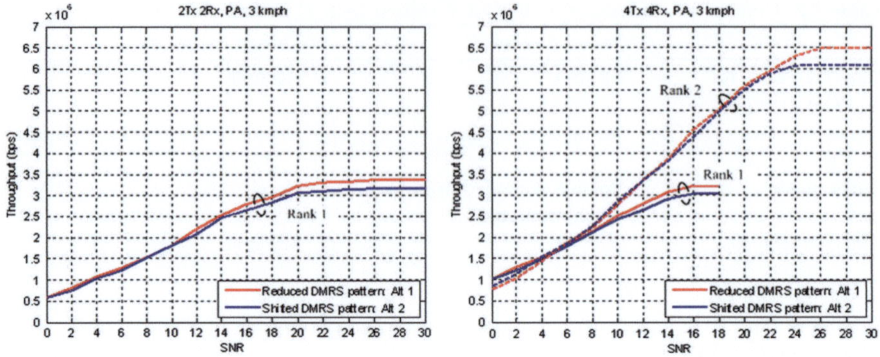

Fig. 4.14 Link level backhaul PDSCH throughput as a function of SNR

The shifted DM-RS pattern (Alt.2) shows slight performance benefit when SNR is low or moderate, i.e., <6 dB for rank = 1, <13 dB for rank = 2. Low SNR prompts the need for more accurate channel estimation. As SNR increases beyond 6 or 13 dB, the reduced DM-RS pattern (Alt 1) shows slightly better performance, mainly due to the smaller overhead. The intuition is that at high SNR region, more accurate channel estimation becomes less crucial as the channel estimation error is already low enough to support the modulation order being used. This principle applies both rank = 1 and rank = 2 MIMO transmissions.

The performances of these two alternatives do not differ much as the better channel estimation is offset by the extra overhead in shifted DMRS (Alt. 2). Also considering the fixed deployment of RN where strong LOS environment in the backhaul and high rank MIMO transmission becomes less likely, RAN1 decided to use reduced DMRS (Alt. 1). In [3], the reduced DMRS is specified as:

> *The reference signal sequence of antenna port* 7, 8, 9 *and* 10 *shall not be mapped to resource elements in the second slot of a PRB pair used for eNB-to-RN transmission when configuration* 1 *in* Table 4.3 *is used.*
>
> Antenna ports 11–14 shall not be used for eNB-to-RN transmission.

Release 10 DMRS can have 12 or 24 REs per PRB pair. Therefore, when a RN's DL grant is in the first slot of a PRB pair, and its PDSCH is the second slot of that PRB pair, the DMRS overhead in the first slot can be either 6 or 12 REs, depending on whether rank = 2 or 4 is used for the PDSCH transmission. That overhead should be known to the RN in order to correctly demodulate the DL grant (R-PDCCH). Three solutions were proposed:

- RN always assumes the maximum overhead;
- RN blindly decodes;
- Higher layer signaling.

As a trade-off between the resource waste in Solution 1 and the blind decoding complexity in Solution 2, Solution 3 was adopted and the DMRS overhead information is derived from the higher-layer parameter for codebook restriction. The exact wording in [3] is

> *For R-PDCCH according to* Sect. 5.6.2 (R-PDCCH formats without cross-interleaving), *if the RN is configured to receive PDSCH data transmissions according to transmission mode* 9, *the RN may assume that the REs for UE-specific reference signals according to the maximum restricted rank are not used for transmission in the first slot of VRB pairs that are used for R-PDCCH transmission, where the higher-layer parameter codebook-SubsetRestriction-r*10 *indicates the maximum restricted rank.*

To ensure the robustness reception of R-PDCCH, R-PDCCH should have the following restrictions:

- Only rank 1 transmission is supported for R-PDCCH per RN, similar to PDCCH, no single user MIMO. This is applied to both CRS and DMRS based R-PDCCH.
- In the DMRS case, antenna port is fixed to be $p = 7$, and the scrambling code ID should be 0, i.e., the specification does not explicitly support multi-user MIMO for R-PDCCH.
- Only QPSK modulation.

The corresponding exact wording in [3] is:

> *If the R-PDCCH is demodulated based on UE-specific reference signals:*
> *Single antenna port; port 7 and* $n_{SCID} = 0$ *is used*
> *If the R-PDCCH is demodulated based on cell-specific reference signals:*
> *If the number of PBCH antenna ports is one:*
> *Single-antenna port; port* 0 *is used*
> *Otherwise Transmit diversity is used*

The above wording is in the context of fallback mode in Transmission Mode 8 and Transmission Mode 9 for RN's PDSCH. In Release 8/9/10 LTE, each transmission mode for PDSCH has a "fallback" mechanism to allow UEs to perform basic radio link set up or reconfiguration. DCI format 1A is used in this fallback mode.

In Release 9, dual-layer beamforming is introduced to enhance the downlink transmission, especially for TDD systems with 8Tx cross-polarization antennas. The corresponding transmission mode is Mode 8, with DCI format 2B. In Release 10, downlink transmission can support up to 8 by 8 MIMO, dynamic switching between single user MIMO and multi-user MIMO, and double codebook feedback in the case 8Tx cross-polarization antennas for FDD. The corresponding transmission mode is Mode 9, with DCI format 2C. DRMS port 7 is used in both Transmission Mode 8 and Transmission Mode 9, however, the fallback mode of DCI format 1A still uses common reference signal (CRS).

Since relay node can be configured with DMRS for R-PDCCH, reusing CRS based DCI format 1A for fallback mode means that the relay node should support both CRS and DMRS, even it is configured for DMRS. To alleviate the burden to always support CRS for R-PDCCH, DMRS based DCI format 1A operation is defined, where port 7 and scrambling ID = 0 are used in the backhaul fallback mode.

4.2.7 Cross-Interleaved and Non Cross-Interleave R-PDCCH

The total resources of R-PDCCHs in a backhaul subframe are configured by higher-layer in a semi-static fashion. R-PDCCH resources can be either frequency distributed or localized. R-PDCCHs of different RNs can be cross-interleaved, similar to Release 8 PDCCH. In cross-interleaved mode, the resource elements in a PRB would contain R-PDCCHs of multiple RNs. Cross-interleaving in the first slots for DL grants and in the second slots for UL grants would be independent, for the purpose of early decoding and efficient resource utilization.

The use cases of cross-interleaved mode are:

- When the backhaul propagation environment is not very favorable, i.e., occasional deep fades are expected either in time and frequency domains, so that interleaving can average out the fading and make R-PDCCH transmission more reliable.
- When the number of RNs per cell is relatively large, multiple R-PDCCHs of different sizes (either because of the different DCI formats, or the aggregation levels) can be aggregated to reduce the fragmentation and improve the resource utilization, i.e., better trunking efficiency.

For cross-interleaved R-PDCCH, it is natural to use only CRS for demodulation since CRS is common to RNs. DMRS seems an overhead burden, rather than beneficial, since DMRS is RN specific. Therefore, [3] specifies:

> *The R-PDCCH according to* Sect. 5.6.3 (R-PDCCH formats with cross-interleaving) *shall be demodulated based on cell-specific reference signals transmitted on one of antenna ports* {0}, {0, 1}, *or* {0, 1, 2, 3}.

Alternatively, R-PDCCH can directly be encoded and mapped to a PRB without being cross-interleaved with other RN's R-PDCCH. This mode in some sense resembles PDSCH resource allocation.

Non cross-interleave mode has the following merits:

- Frequency selective gain. Without interleaving, a RN's R-PDCCH has the freedom to be transmitted in certain PRBs that would experience up fades.
- Precoding gain. RN specific precoding or beamforming can effectively be applied to each RN to boost te received SNR for R-PDCCH
- Operation flexibility. Each RN's R-PDCCH is independently encoded and mapped to a resource that is suitable.

DMRS is obviously suitable for non cross-interleaved R-PDCCH so that it can fully benefit from the precoding gain. CRS, while less helpful for precoding (since RN does not know TPMI without decoding DCI in the first place), is still attractive for its less overhead. Therefore, [3] specifies that

The R-PDCCH according to Sect. 5.6.2 (R-PDCCH formats without cross-interleaving) *shall be demodulated based on cell-specific reference signals transmitted on one of antenna ports* {0}, {0, 1}, *or* {0, 1, 2, 3}, *or based on UE-specific reference signals transmitted on antenna port* 7 *assuming that NSCID* = 0, *the type of reference signals is configured by higher layers.*

Both cross-interleaved and non cross-interleaved R-PDCCH modes are supported in Release 10 relay to accommodate different deployment scenarios. In the following, more details of R-PDCCH specifications are described. Let us first look at cross-interleaved R-PDCCH.

As the purpose of cross-interleaved mode is to make R-PDCCH detection more robust, the mapping from R-PDCCHs to the PRBs should be frequency diverse. There are basically two approaches to design such frequency diverse resource mapping:

1. To introduce a new mapping and search space design.
2. To reuse Release 8 resource allocation type and search space design.

An example is given [12] on how to design a new resource mapping to achieve frequency diversity and to fully utilize the resource, while maintaining reasonable blind decoding complexity. Even though a relay node potentially has more processing power to carry out more blind decodes, big deviation from Release 8 PDCCH blind decoding complexity is still not desirable. With the consideration of false alarm rate and the significant more work to be done in RAN4 to test R-PDCCH, the number of blind decodes for R-PDCCH was decided to be around 44, similar to that of Release 8 PDCCH.

A group of PRB subsets can be defined within the semi-statically assigned PRBs for R-PDCCH transmission as Fig. 4.15 shows. The system bandwidth is 5 MHz that has 25 PRBs, among which 8 PRBs are for R-PDCCHs, i.e., {0, 3, 6, 9, 13, 16, 19, 22}, defined as Subset 1. Similarly, Subset 2 and Subset 3 can be defined as: Subset 2, with 4 PRBs, indices {0, 3, 13, 16}; Subset 3, with 2 PRBs, indices {0, 13}.

A relay node would start blind decoding by checking Subset 3 first. If no R-PDCCH for this RN is detected, it then tries to decode R-PDCCH in Subset 2. If no R-PDCCH is found, the RN would continue decoding in Subset 3.

The number of blind decodes can be calculated as follows. Consider the case of detection of DL grant in the first slot and assume that control channel element (CCE) for R-PDDCH can fit into one PRB, the above three subsets with 8, 4, 2 PRBs constitute a common search space for a group of relay nodes. So for 8 PRB common search space, 8, 4, 2 and one blind decodes are needed for 1, 2, 4 and 8 CCE aggregation levels, respectively. For 4 PRB common search space, 4, 2 and one blind decodes are needed for 1, 2, and 4 CCE aggregation levels, respectively. For 2 PRB common search space, 2 and one blind decodes are needed for 1 or 2 CCE aggregation levels, respectively. The total blind decodes is 25. If downlink fall back mode, i.e., DCI format 1A is configured in each transmission mode for the relay, the total would be 50. Similar for the uplink grants in the second slot, if the resource mapping follows the same pattern, 25 blind decodes would be needed.

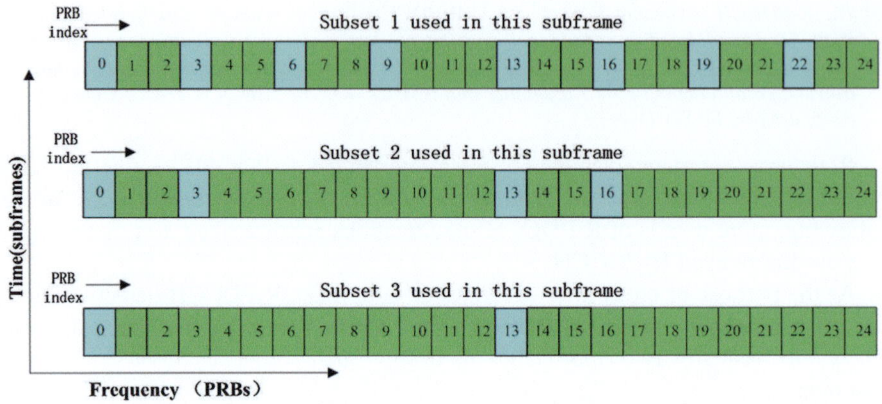

Fig. 4.15 PRB subsets for R-PDCCH resources to support frequency diversity

So the maximum number of blind decodes of a relay node in order to detect the DL grant and UL grant is 75 in this case.

In this example of resource mapping and search space design, the number of PRBs actually used for R-PDCCH can vary dynamically and the rest of resources can be used for PDSCH transmission, thereby improving the resource utilization especially for cross-interleaved R-PDCCH. However, such resource utilization benefit comes with the price of increased number of blind decodes, which seems undesirable. Also, a DL grant or UL grant of CCE aggregation level 1 may not nicely fit one slot of PRB. The PRB subsets shown in Fig. 4.15 may not generally be applicable for various DCI payload sizes, different operating bandwidth, and different relay frame timings. Therefore, *the prevailing view in RAN*1 *was to reuse the resource mapping and search space design of Release* 8 *PDCCH for cross-interleaved R-PDCCH.*

Release 8 PDCCH search space is a logic domain concept, i.e., its definition has nothing to do the actual mapping to the physical resources or PRBs. This makes it possible to separately design and optimize the search space, without worrying about the actual physical resources. The same principle can be used for R-PDCCH. In fact, the different resource allocation types for PDSCH in Release 8 already support both localized and distributed resource mapping from virtual resource blocks (VRB) to PRB. For example, with localized virtual resource block (LVRB), VRBs are mapped directly to PRBs, i.e., $n_{\mathrm{PRB}} = n_{\mathrm{VRB}}$. By using distributed virtual resource block (DVRB), two VRBs in a VRB pair are mapped to different PRBs in two slots of a subframe. With this already defined mechanism, the resources for R-PDCCH can be configured in terms of set of VRBs, which is reflected in [3] as

> *The relay node shall monitor the set of configured VRBs in the first slot for an R-PDCCH containing a downlink assignment and it shall monitor the set of configured VRBs in the second slot for an R-PDCCH containing an uplink grant.*

And this applies to without cross-interleaving R-PDCCH.

With the good channel quality of the backhaul, whether to support 8 CCE aggregation for cross-interleaved R-PDCCH became a relevant question. The simulation in [13] shows that for DCI format 1, one CCE aggregation has roughly 93 % of chance to be used, 2 CCE aggregation has roughly 5 % probability, 4 CCE aggregation has about 1 %, and 8 CCE aggregation has about 1 % probability. For DCI format 2, the distribution is slightly different, roughly {87, 9.5, 2, 1.5 %} for CCE aggregation levels 1, 2, 4 and 8, respectively. It is noted that the probability of using 8 CCE aggregation level varies significantly for different deployments and can be up to nearly 3 % in Case 3 rural/suburban propagation scenario for DCI format 2. Therefore, in order to make sure reliable detection of R-PDCCH in cross-interleaved mode, *CCE aggregation level* 8 *is supported in Release* 10 *relay*.

Up to now, from the search space aspect, cross-interleaved R-PDCCH can reuse most designs of Release 8 PDCCH, with only two exceptions:

1. Two separate search spaces are needed for DL grants in the first slots and for UL grants in the second slots, respectively.
2. The total number of CCEs is configured by higher-layers, rather than derived from PCFICH, operating bandwidth, etc.

The exact wording in [3] is:

> *The set of CCEs corresponding to an R-PDCCH candidate m of the search space* $S_{n,j}^{(\Lambda)}$ *in slot* $j \in \{0, 1\}$ *of subframe n is given by* $\Lambda \cdot \left\{ (Y_n + m) \bmod \left\lfloor N_{\mathrm{CCE},j}^{\mathrm{R-PDCCH}} / \Lambda \right\rfloor \right\} + i$ *where* $i = 0, 1, \ldots, \Lambda - 1$, $m = 0, 1, \ldots, M(\Lambda) - 1$, *and* $N_{\mathrm{CCE},j}^{\mathrm{R-PDCCH}}$ *is the total number of CCEs in the set of RBs configured for potential R-PDCCH transmission.*
>
> *The relay node shall only monitor one RN-specific search space according to the UE-specific search space in* [5] *at each of the aggregation levels* $\Lambda \in \{1, 2, 4, 8\}$

The last detail of cross-interleaved R-PDCCH specification is the definition of resource element group (REG). In Release 8 PDCCH, one REG occupies only one OFDM symbol. Since there are only CRS in PDCCH region, REG layout is very regular and evenly distributed. However, R-PDCCH resides in PDSCH region. So the presence of DMRS and CSI-RS would lead to rather complicated REG patterns for R-PDCCH. With these considerations, it was proposed in [14] to allow one REG spanning over two consecutive OFDM symbols as Fig. 4.16 shows. Note that there is no CSI-RS present in these subframes. The proposed REG definition has the merit in reducing REG fragmentation and improving the resource element utilization for R-PDCCH. However, it may not be efficient when CSI-RS is present. A more important concern is that such two OFDM symbol REG definition significantly deviates from Release 8 REG definition and would require very different implementation of R-PDCCH reception. Therefore, *the new REG definition that spans two consecutive OFDM symbols was not agreed.*

On the other hand, it is understandable that sticking to Release 8 REG principle would incur significant resource waste and complicated REG patterns if we want

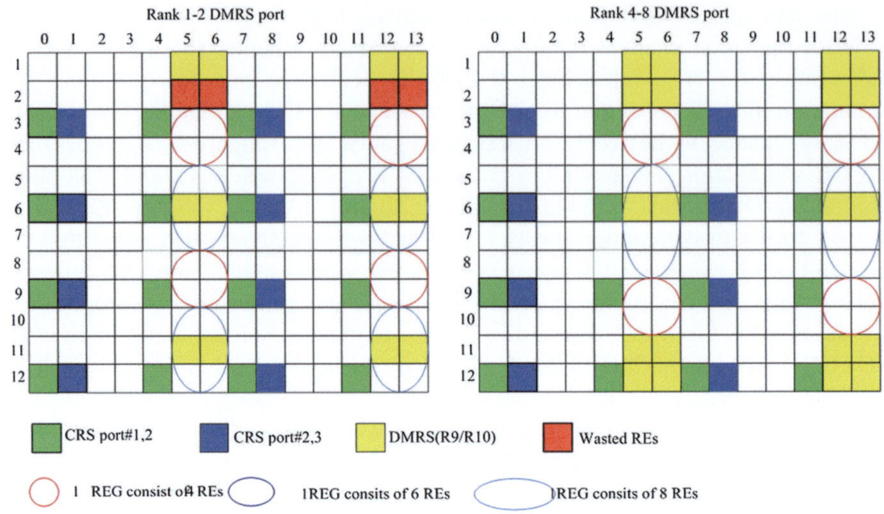

Fig. 4.16 Two OFDM symbol REG definition proposed for cross-interleaved R-PDCCH, not adopted

to accommodate both CSI-RS and DMRS in a PRB pair. Some trade-off has to be made: we may avoid either CSI-RS or DMRS.

CSI-RS is cell specific and would be present across the entire bandwidth if it is configured in a subframe. Therefore, it is difficult to avoid CSI-RS completely.

DMRS is RN specific. Since cross-interleaved R-PDCCH is only modulated by CRS. DMRS would not be present when both the first slot and the second slot in a PRB pair carry R-PDCCH. It only becomes an issue if the first slot of a PRB pair carries DL grant transmission and the second slot of this PRB pair carries PDSCH modulated by DMRS. However, many companies believed that such situation can be avoided by proper scheduling. With this burden lifted off, REG definition becomes cleaner as illustrated in Fig. 4.17.

The actual wording in the specification [3] is as follows:

- an REG is composed out of 4 consecutively available REs in one OFDM symbol in a PRB counted in ascending order of subcarriers, where an RE is assumed to be unavailable with respect to mapping the R-PDCCH in the following cases:
- if it is used for the transmission of cell-specific reference signals
- if the cell-specific reference signals are configured to be transmitted only on antenna port 0, it shall be assumed that REs for transmission of cell-specific reference signals on antenna port 1 are unavailable for an REG
- if zero power or non-zero power CSI-RS occurs in any resource element of an eight-port CSI-RS configuration of Table 6.10.5.2-1 of [3], it shall be assumed that all eight resource elements corresponding to the eight-port CSI-RS configuration are unavailable for an REG
- UE-specific reference signals are not mapped onto PRB pairs used for the transmission of R-PDCCH with cross-interleaving
- for the purpose of REG-to-RE mapping

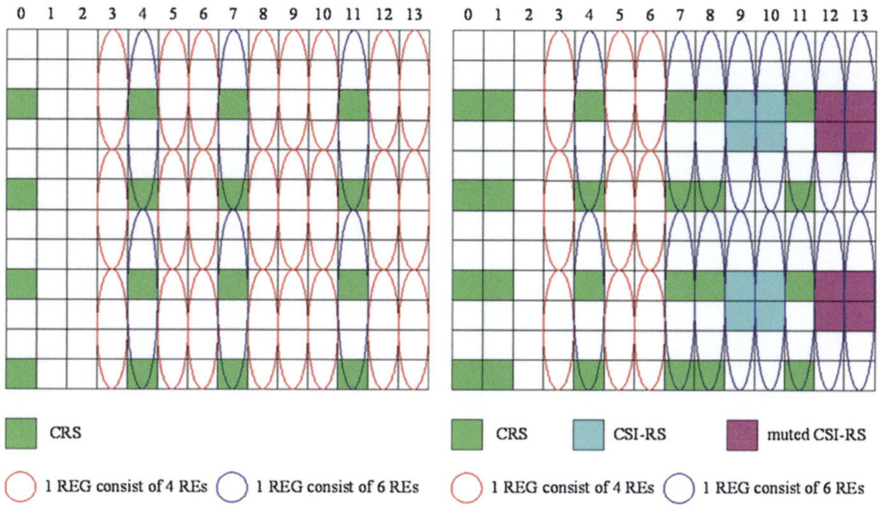

Fig. 4.17 Two examples of agreed REG mapping for cross-interleaved R-PDCCH, adopted

- the downlink system bandwidth shall be determined as $N_{\mathrm{VRB}}^{\mathrm{R-PDCCH}}$
- the time-domain index l' shall be initialized with the start symbol index given in Tables 5.4.1 and 5.4.2 respectively and l' shall be increased if $l' \leq L$, where L corresponds to the end symbol index given in Tables 5.4.1 and 5.4.2 respectively.

Now let us look at R-PDCCHs without cross-interleaving. First of all, it is reasonable to set the granularity of R-PDCCH for aggregation level 1 to be one slot of PRB pair. The first slot of backhaul subframe would have about 36–44 resource elements for R-PDCCH which is comparable to one CCE.

It should be noted that although one CCE size is comparable to that of a slot of PRB, they have different meanings. CCE is defined for PDCCH and cross-interleaved R-PDCCH. CCE is purely a logic concept and does not correspond to any particular PRB or even to specific set of resource elements.

There is a key parameter in Release 8 PDCCH search space design, the Hashing function which changes randomly slot by slot, but known to the receiver beforehand. The Hashing function was defined with two purposes: (1) to reduce the persistent blocking when search spaces collide; (2) to randomize the resources used by PDCCH or R-PDCCH and improve the robustness of PDCCH decoding. Such randomization of search space does not seem necessary for non-interleaved R-PDCCH that relies on frequency selectivity and precoding to boost the performance or capacity. Therefore, *RAN*1 *decided not to reuse Release* 8 *Hashing function for search space of non-interleaved R-PDCCH. And the search space is defined at resource block granularity.*

The search space for the first slots and the second slots can be the same since it would fully benefit the frequency selective gain and precoding gain, either when both slots carry DL/UL grants, or the second slot is for PDSCH transmission for this RN.

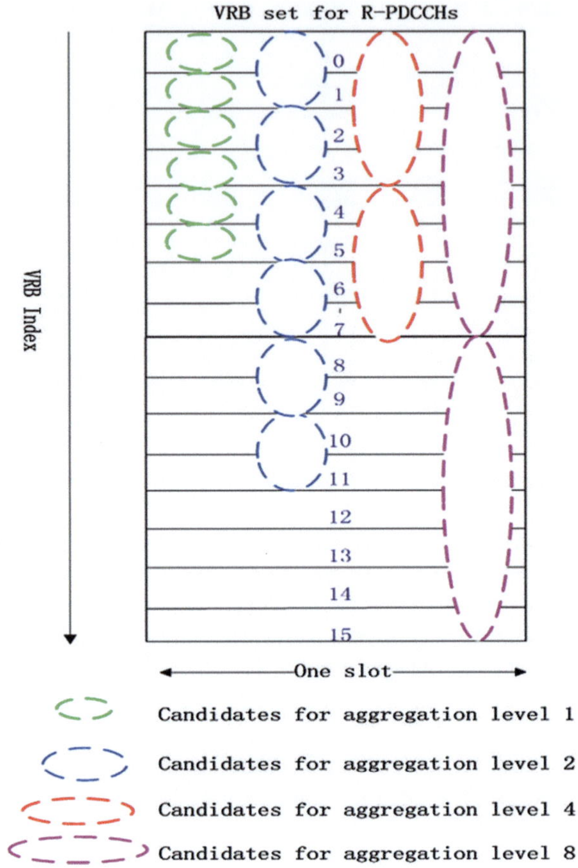

Fig. 4.18 An example of R-PDCCH search space candidates for non-interleaved R-PDCCH

In [15], an example (as Fig. 4.18) is used to illustrate the search space design when 8 VRBs are higher-layer configured for non-interleaved R-PDCCH transmission, Four aggregation levels are used: {1, 2, 4, 8} with number of candidates {6, 3, 1, 1}, respectively.

Ideally, the number of candidates per aggregation level could be further optimized to reflect the typical deployment of relay backhaul. However, due to the limited time for extensive performance study, numbers in Release 8 PDCCH are reused.

The exact wording in the specification [3] is:

The same set of VRBs is configured for a potential R-PDCCH in the first and in the second slot.

In each slot, an R-PDCCH candidate $m = 0, 1, \ldots, M(\Lambda) - 1$ *at aggregation level* Λ *comprises VRB numbered with* $n_{\mathrm{VRB}}^{\mathrm{R-PDCCH}} = (\Lambda \cdot m + i) \bmod N_{\mathrm{VRB}}^{\mathrm{R-PDCCH}}$, *where* $i = 0, 1, \ldots, \Lambda - 1$ *and where* $M(\Lambda)$ *is* the same as Release 8 PDCCH.

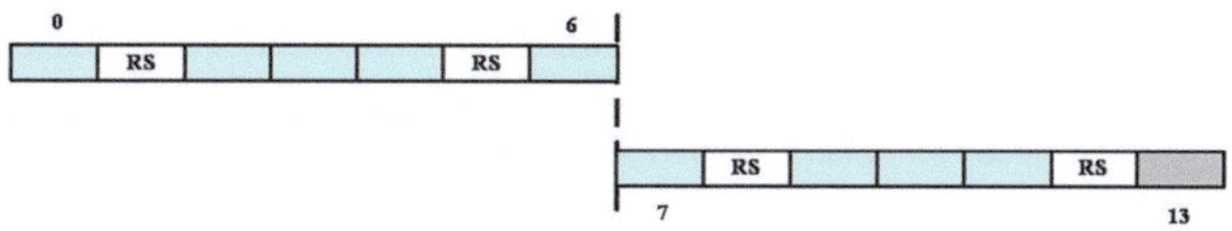

Fig. 4.19 Structure of shortened PUCCH format 2/2a/2b

4.2.8 PUCCH

Release 8 defines an implicit mapping between CCE index and the ACK/NAK resource. In principle, dynamic ACK/NAK for Un can share the same resources as UE, provided that scheduler ensures no collision on the dynamic ACK/NAK resource between RN and UE. This, however, adds scheduler restriction. In addition, the timing between PDSCH transmission to ACK/NAK feedback can be different for RN and UE. Hence, dynamic ACK/NAK resource sharing between RN and UE leads to overly complicated schedulers. Therefore, it makes sense that the set of dynamic ACK/NAK resources for RNs is different from that of UEs. Considering that the number of RNs scheduled in the same DL backhaul subframe is limited, RRC signaling can be used to configure the ACK/NAK resource on RN specific base [16], more specifically via $n_{\mathrm{PUCCH}}^{(1,p)}$ and $n_{\mathrm{PUCCH},i}^{(1)}$.

Another topic been discussed for PUCCH is whether the resource mapping can be improved in shortened format 2/2a/2b. As discussed earlier, backhaul uplink subframe may have only 13 SC-FDMA symbols where the shortened format 2/2a/2b should be used, as illustrated in Fig. 4.19.

It has been observed that in Reed-Muller code (20, A) as copied in Table 4.7, the first ten elements of sixth column $M_{i,5}$ are all zeros, meaning that this column does not contribute to the frequency hopping diversity, since the frequency hopping occurs at the slot boundary, translated into the tenth bits. Certain loss of frequency diversity is expected when the number of payload bits is greater than 5, i.e., $A > 5$.

There was a speculation on whether such flaw can be mitigated by shifting the resource mapping, or puncturing the first few rows, instead of the last few rows.

To verify that speculation, the link level simulation of those two choices was carried out:

1. Puncturing the first two rows.
2. Puncturing the last two rows.

EPA channel is assumed whose delay spread is relatively small, reflecting the backhaul channel characteristics. The required SNRs in dBs for block error rate of 1 % are compared in Table 4.8. Choice 1 shows better performance than Choice 2 for payload more than 4 bits. Similar observation is seen in [18].

Choice 1 requires specification change. The change can be very minor, i.e., adding a shift when ACK/NACK bits are mapped to SC-FDMA symbols, without touching Reed-Muller (20, A) code. In the end, given the limited interest in changing the spec, *no agreement was reached on this issue.*

Table 4.7 Basic sequence of (20, A) code [17]

i	$M_{i,0}$	$M_{i,1}$	$M_{i,2}$	$M_{i,3}$	$M_{i,4}$	$M_{i,5}$	$M_{i,6}$	$M_{i,7}$	$M_{i,8}$	$M_{i,9}$	$M_{i,10}$	$M_{i,11}$	$M_{i,12}$
0	1	1	0	0	0	0	0	0	0	0	1	1	0
1	1	1	1	0	0	0	0	0	0	1	1	1	0
2	1	0	0	1	0	0	1	0	1	1	1	1	1
3	1	0	1	1	0	0	0	0	1	0	1	1	1
4	1	1	1	1	0	0	0	1	0	0	1	1	1
5	1	1	0	0	1	0	1	1	1	0	1	1	1
6	1	0	1	0	1	0	1	0	1	1	1	1	1
7	1	0	0	1	1	0	0	1	1	0	1	1	1
8	1	1	0	1	1	0	0	1	0	1	1	1	1
9	1	0	1	1	1	0	1	0	0	1	1	1	1
10	1	0	1	0	0	1	1	1	0	1	1	1	1
11	1	1	1	0	0	1	1	0	1	0	1	1	1
12	1	0	0	1	0	1	0	1	1	1	1	1	1
13	1	1	0	1	0	1	0	1	0	1	1	1	1
14	1	0	0	0	1	1	0	1	0	0	1	0	1
15	1	1	0	0	1	1	1	1	0	1	1	0	1
16	1	1	1	0	1	1	1	0	0	1	0	1	1
17	1	0	0	1	1	1	0	0	1	0	0	1	1
18	1	1	0	1	1	1	1	1	0	0	0	0	0
19	1	0	0	0	0	1	1	0	0	0	0	0	0

Table 4.8 Required SNR for PUCCH format 2/2a/2b for EPA channel

	4 bits	6 bits	8 bits	10 bits	11 bits
Solution 1	−6.34	−4.71	−3.52	−2.13	−1.15
Solution 2	−6.44	−4.08	−3.25	−1.93	−0.6

4.3 Backhaul Subframe Configuration and HARQ Timing

In LTE, a radio frame (10 ms) contains 10 equally divided subframes. For TDD, different numbers UL and DL subframes can be configured, to serve asymmetric traffic patterns. Note that subframes #0, #4, #5 and #9 in FDD and #0, #1, #5 and #6 in TDD carry system information, synchronization channels, paging channels, etc. Those subframes should be visible to Release 8 UEs at all times and cannot be configured as MBSFN subframes in RN cells. As agreed during the study item phase of Release 10 relay, DL backhaul subframes should be semi-statically and explicitly configured by RRC signaling for both FDD and TDD systems.

In addition to the MBSFN subframe constraint, backhaul subframe allocation should also consider HARQ timing in both backhaul link and access link. Backhaul subframe configuration and HARQ are closely related. There are a few challenges during the specification of Release 10 RN:

- UL HARQ timing in FDD follows synchronous HARQ with 8 ms round trip time. Yet, MBSFN subframe is of 10 ms period. So HARQ timing and MBSFN subframe occurrence are not well matched.
- In TDD, UL grant timing and UL ACK/NACK timing do not always fit each other in some UL-DL configurations.
- In TDD, some UL-DL configurations see extreme imbalance in terms of number of UL and DL subframes in a radio frame.

4.3.1 FDD systems

In Release 8 FDD, the uplink HARQ is synchronous and the HARQ timing associated with uplink transmission is relatively simple and fixed, compared to the case in TDD. Three HARQ timing relationships are important to the discussion of uplink backhaul subframe allocation:

1. PDSCH and the corresponding UL ACK/NACK.
2. UL grant (sent in PDCCH) and the corresponding PUSCH.
3. PUSCH and the corresponding DL ACK/NACK.

The response time is 4 ms in all three HARQ timings shown above. That is, UE would take 4 ms to respond eNB whether PDSCH is received, by sending ACK/NACK on UL. UE needs 4 ms to decode an UL grant target to it, and send PUSCH. eNB takes 4 ms to decode PUSCH and respond UE via DL ACK/NACK. Uplink HARQ is synchronous in the sense that UL grant and DL ACK/NACK occur in the same modulo 8 subframes, i.e., round trip time (RTT) is fixed to 8 ms.

Since the first two timings match each other, a nature way is to reuse Release 8 PDSCH to UL ACK/NACK timing and UL grant to PUSCH timing, and derive UL backhaul subframes implicitly from DL backhaul subframe allocation. That is: if a subframe indexed by k is configured as backhaul downlink subframe, the subframe 4 ms later, i.e., indexed by $k + 4$, would be implicitly configured for backhaul uplink subframe which may carry UL ACK/NACK (or PUSCH) to respond PDSCH (or UL grant) in downlink subframe indexed by k. In [3], it is specified that

> *A subframe n is configured for RN-to-eNB transmission if subframe $n - 4$ is configured for eNB-to-RN transmission.*

While the above two Release 8 HARQ timings can easily be maintained by implicit configuration of UL backhaul subframes, the third timing, DL ACK/NACK to PUSCH, may not always be fulfilled. Due to the 10 ms period of MBSFN subframes, DL ACK/NACK sometimes may not be sent 4 ms after the PUSCH. Making it less than 4 ms is certainly highly undesirable as that would require more processing power at eNB. So the timing between DL ACK/NACK and PUSCH has to be variable, i.e., the index of downlink backhaul subframe should be $\geq k + 4$, for the corresponding uplink backhaul subframe indexed by k.

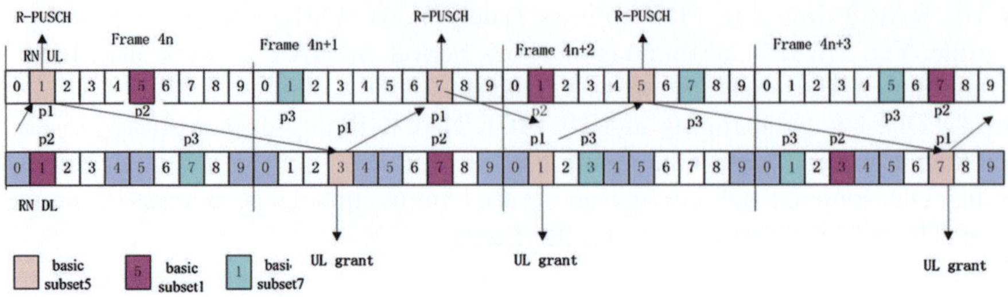

Fig. 4.20 An example of backhaul HARQ timing with RTT of 8 or 16 ms

Because of this, the round trip time for backhaul uplink transmission would be larger than 8 ms.

There are three timing options for DL ACK/NACK to respond backhaul PUSCH:

Option 1: Round trip time (RTT) fixed to 8 or 16 ms

Least specification change is expected for Option 1. DL ACK/NACK would occur 4 or 12 ms after the transmission of PUSCH, as illustrated in Fig. 4.20. Least effort is needed to modify the scheduler as the absence of ACK at 8 ms can be treated as if the ACK was not received, a similar problem that can occur even in Release 8.

Since both backhaul link of Option 1 and the access link have the same round trip time of 8 ms, the number of collisions them can be minimized by assigning the subframes associated to same access link uplink HARQ process as in the backhaul link.

In Option 1, the HARQ process number is obvious, just by numbering sequentially over the period of backhaul subframe allocation according to the fixed RTT. Release 8 design of uplink HARQ process identification can be reused, and the number of backhaul uplink HARQ process equals to the number of configured basic subsets to backhaul subframes. Therefore, at most 8 HARQ process IDs can be allocated to backhaul uplink subframes. RN maps one basic subset to one uplink HARQ process ID.

In the example [19] shown in Fig. 4.20, subframes 1, 5, 7 are configured for backhaul downlink. Therefore, the number of backhaul uplink HARQ processes is 3. The basic subset 5 can be allocated as uplink HARQ process ID = 1, basic subset 1 to uplink HARQ process ID = 2, basic subset 7 to uplink HARQ process ID = 3. The UL grants in DL subframes #13, 21, 37, and PUSCH in UL subframes # 1, 17, 25 belong to the uplink HARQ process ID = 1.

Option 2: Round trip time (RTT) fixed to 10 ms

Figure 4.21 shows an example of RTT fixed to 10 ms. The most significant merit of Option 2 is that the HARQ patterns perfectly match 10 ms period of backhaul subframe allocation. Indirectly, Option 2 bears certain similarity with

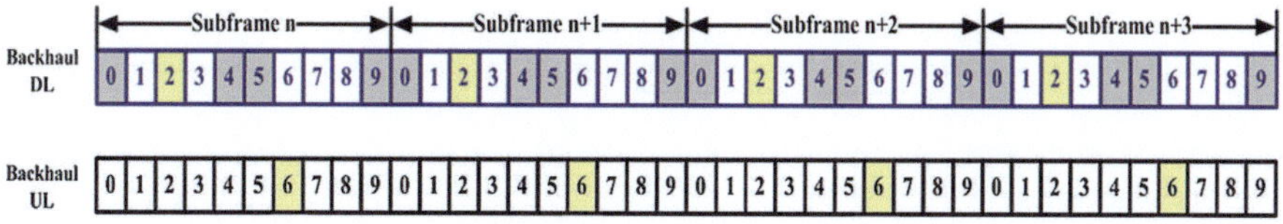

Fig. 4.21 An example of fixed 10 ms HARQ RTT on backhaul uplink

TDD whose HARQ timing often has to fit the DL/UL subframe allocation. Also in TDD, it has been agreed to reuse Release 8 TDD fixed 10 ms RTT.

For 10 ms fixed RTT, there are up to 6 HARQ processes, and backhaul uplink subframes are implicitly derived from backhaul downlink subframes in 10 ms interval. Because of this, the HARQ timing in Option 2 is drastically different from Release 8 uplink where the RTT is 8 ms. Note that since RTT of 8 ms is used in access link (for Release 8 backward compatibility purpose), the HARQ of 10 ms RTT in the backhaul uplink would lead to frequent collisions with the HARQ of 8 ms RTT in the access uplink. While the HARQ collision issue can be mitigated by certain implementations, the issue cannot be effectively mitigated, thus making Option 2 less attractive.

Option 3: Variable round trip time (RTT $\geq$ 8 ms)

Option 3 is motivated by minimizing the HARQ latency. It shares some design philosophy with Option 1 in the sense that RTT of 8 ms should be kept as much as possible, otherwise using different values (>8 ms). In Option 3, the first available subframe associated to same HARQ process is used, which leads to the same issue as Option 2 of fixed 10 ms RTT: 8 ms periodicity cannot be maintained. Therefore, substantial standardization effort is needed, which involves two steps:

Step 1: Determining the number of HARQ processes

The minimum HARQ RTT, i.e., the time from a HARQ transmission attempt until the earliest occurrence of a retransmission is 8 ms for the backhaul link. Consequently, 8 HARQ processes are required to be able to continuously transmit, if all subframes can be used for transmission. In the case of in-band relaying, less subframes are available in an 8 ms interval. Therefore, less HARQ processes are needed. The number of uplink HARQ processes is equal to the number of configured backhaul subframes in an 8 ms interval. This number could be determined, for instance, by a window of 8 ms RTT which slides over the backhaul subframe configuration, as illustrated in Fig. 4.22 with 40 ms period [20]. Here 6 subframes are configured as backhaul subframes. Non-MBSFN subframes are in bluish grey. In this configuration there are at most 3 backhaul subframes within the sliding window. Therefore, 3 HARQ processes are required for this configuration.

Step 2: Association of subframe number and HARQ process ID

Having determined the number of HARQ processes, Un subframes can be associated to HARQ process IDs, e.g., by means of sequential numbering using the modulo operation. Figure 4.23 shows the example configuration of Fig. 4.22 where HARQ process IDs are associated to Un subframes.

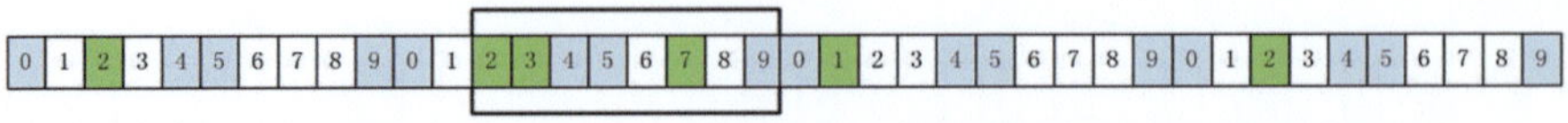

Fig. 4.22 Variable RTT in a 40 ms Un subframe configuration period (numbers referred to the subframe numbering)

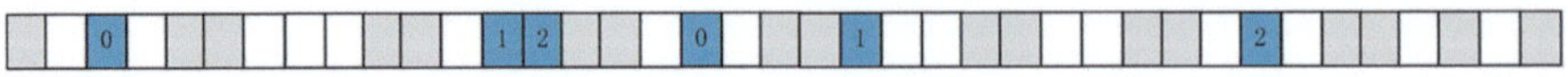

Fig. 4.23 Association of subframe number and HARQ process ID (numbers referred to the subframe numbering)

Note that the described process of numbering (i.e., determining the minimum number of HARQ processes and associating HARQ process IDs to subframes) needs to be performed only when backhaul subframes are initially configured or re-configured. During normal operations, the numbering is constant.

In terms of HARQ latency, the intuition is that Option 3 should perform better than the two other options. Option 1 can minimize latency for some packets, but has to increase latency for other retransmissions. Option 2 should sit in the middle, with a uniform but larger than minimum 8 ms value. However, the simulation in [21] shows that their differences in HARQ latency are actually not significant, as depicted in Fig. 4.24. While Option 3 seems consistently slightly better than Option 1, for different numbers (between 3 and 24) of configured backhaul uplink subframes, Option 2 has least delays when the number of configured uplink subframes is small. In the simulation, two-hop latency includes the HARQ delays in both backhaul link and access link. The initial target error rate is 10 % in both links.

In option 3, backhaul link and access link HARQ collisions may sometimes be comparable to Option 1, as illustrated in Fig. 4.24. For Option 1, collisions are minimized by using proper HARQ processes between backhaul and access links, i.e., assigning 3 subframes associated to HARQ process ID = 5 for backhaul link. Hence, only one process of Uu link is impacted by backhaul link HARQ. In Option 2, four access link uplink HARQ processes (1, 3, 5, 7) are impacted when four backhaul subframes (associated 4 different HARQ processes) are assigned. In Option 3, four access uplink HARQ processes (1, 2, 3, 5) are impacted when 4 backhaul subframes are assigned. This example shows that in Option 3, the collisions of HARQ processes in backhaul and access links can be more frequent and harder to avoid than the case of Option 1.

Option 3, variable RTT choice, was agreed in RAN1 as it was generally believed to result in shorter average RTT, compared to Option 1 of 8/16 ms fixed RTT. The issue of backhaul and access link HARQ collision was believed less severe in Option 3 than in Option 2, 10 ms fixed RTT.

Backhaul subframes are configured by an 8-bit bitmap, i.e., 256 combinations. Over the next 32 subframes, the 8-bit map is repeated 4 times, excluding those MBSFN subframes. Then we get 40 ms long backhaul subframe pattern. By using the procedure shown in Fig. 4.22, we can calculate the number of uplink HARQ

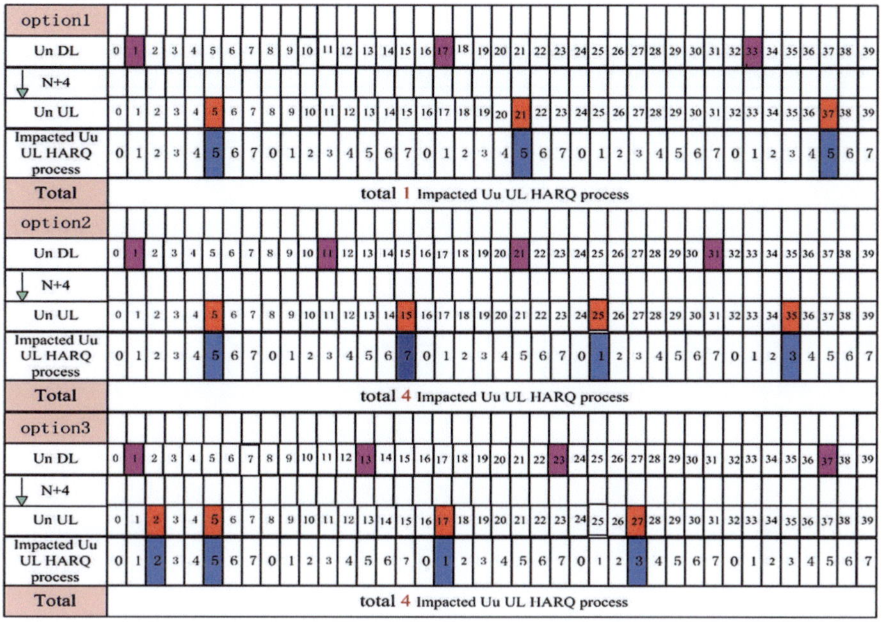

Fig. 4.24 Example of Access link uplink HARQ processes impacted by backhaul link HARQ

processes. The calculation is quite straightforward once the backhaul subframes are configured. And it is only needed during the initial subframe configuration. Therefore, a table as seen in Table 4.9 is specified in [3] where all possible FDD backhaul configurations are numerated in the decimal equivalent of 8-bit bitmap. The results of the calculations for each configuration, the number of uplink HARQ processes, are listed, in the right column. So the first step of Option 1 is specified in [3] as

> *For frame structure type* 1 *the number of HARQ processes is determined by the decimal equivalent of the binary number representing the* 8*-bit bitmap of the parameters SubframeConfigurationFDD as given by* Table 4.10.

The second step of Option 3 is specified in [3] as

> *HARQ processes are sequentially assigned to subframes configured for RN-to-eNB transmission.*

It was proposed to modify the number of uplink HARQ processes to have a single-value RTT (of 20 ms) to solve the problem of semi-persistent scheduling (SPS)-backhaul collisions. The discussion concluded without agreement due to the following considerations. First of all, a scenario, where SPS traffic (i.e. VoIP) occupies the resources of entire subframes is very unlikely. VoIP applications generally do not generate lots of data. Second, if there is such a large demand for SPS data, the eNB could configure SPS subframes which do not collide with backhaul subframes. Third, even if the configuration of SPS- and backhaul subframes overlap, a collision of actual SPS- and backhaul allocations can be avoided by scheduling.

Table 4.9 Latency in ms of Un and Uu links combined, for UL HARQ retransmission probability of 10 % in both links

Un UL subframe number in 40 ms	Option1	Option2	Option3
3	16.6	14.8	16.8
4	15.2	14.5	15.2
5	14.8	14.0	14.8
6	13.5	13.6	13.5
7	13.1	12.5	13.0
8	12.9	12.3	12.8
9	12.6	12.1	12.5
10	12.4	12.0	12.3
11	12.2	11.9	12.0
12	12.0	11.8	11.7
13	11.8	11.7	11.2
14	12.3	11.8	12.1
15	12.4	12.1	12.2
16	12.4	12.3	12.3
17	12.4	12.5	12.3
18	12.5	12.7	12.3
19	12.5	13.2	12.4
20	12.6	14.0	12.5
21	12.7	13.3	12.6

4.3.2 TDD Systems

Release 8 TDD supports 7 UL-DL subframe configurations [22] as Table 4.11 shows.

In UL-DL configuration 0, no other subframe, except for special (S) subframes, can be configured for downlink backhaul as subframes {0, 1, 5, 6} have to be used for access link.

In UL-DL configure 5, excluding S subframes, there is only one UL subframe which can be used for either backhaul link or access link, but not both.

Hence, there is a motivation to explore the usage of special subframes in backhaul link and access link [23]. Special subframe scheme has two unique merits: (1) special subframe scheme could support all TDD configurations including TDD configuration 0 and TDD configuration 5; (2) special subframe scheme has no impact on uplink HARQ process. So potentially, special subframe scheme could be used as a general scheme of access-backhaul partitioning for all TDD configurations, or at least as a complementary scheme for baseline solution so that TDD configuration 0, 5 and 6 could also be supported in Release 10 relay.

It is noted that there are several configurations with long GP length, e.g. special subframe configuration 0 and 5 for normal CP case, which could possibly be used for backhaul transmission, either for downlink backhaul transmission, or for uplink backhaul transmission, or even multiplex downlink backhaul transmission and

Table 4.10 Number of HARQ processes corresponding to a subframe configuration in FDD

Decimal equivalent of SubframeConfigurationFDD	Number of uplink HARQ processes
1, 2, 4, 8, 16, 32, 64, 128	1
3, 5, 6, 9, 10, 12, 17, 18, 20, 24,33, 34, 36, 40, 48, 65, 66, 68, 72, 80, 96, 129, 130, 132, 136, 144, 160, 192	2
7, 11, 13, 14, 19, 21, 22, 25, 26, 28, 35, 37, 38, 41, 42, 44, 49, 50, 52, 56, 67, 69, 70, 73, 74, 76, 81, 82, 84, 85, 88, 97, 98, 100, 104, 112, 131, 133, 134, 137, 138, 140, 145, 146, 148, 152, 161, 162, 164, 168, 170, 176, 193, 194, 196, 200, 208, 224	3
15, 23, 27, 29, 30, 39, 43, 45, 46, 51, 53, 54, 57, 58, 60, 71, 75, 77, 78, 83, 86, 87, 89, 90, 91, 92, 93, 99, 101, 102, 105, 106, 107, 108, 109, 113, 114, 116, 117, 120, 135, 139, 141, 142, 147, 149, 150, 153, 154, 156, 163, 165, 166, 169, 171, 172, 173, 174, 177, 178, 180, 181, 182, 184, 186, 195, 197, 198, 201, 202, 204, 209, 210, 212, 213, 214, 216, 218, 225, 226, 228, 232, 234, 240	4
31, 47, 55, 59, 61, 62, 79, 94, 95, 103, 110, 111, 115, 118, 119, 121, 122, 123, 124, 125, 143, 151, 155, 157, 158, 167, 175, 179, 183, 185, 187, 188, 189, 190, 199, 203, 205, 206, 211, 215, 217, 219, 220, 221, 222, 227, 229, 230, 233, 235, 236, 237, 238, 241, 242, 244, 245, 246, 248, 250	5
63, 126, 127, 159, 191, 207, 223, 231, 239, 243, 247, 249, 251, 252, 253, 254, 255	6

Table 4.11 UL-DL subframe configurations in TDD

UL-DL configuration	Subframe number									
	0	1	2	3	4	5	6	7	8	9
0	D	S	U	U	U	D	S	U	U	U
1	D	S	U	U	D	D	S	U	U	D
2	D	S	U	D	D	D	S	U	D	D
3	D	S	U	U	U	D	D	D	D	D
4	D	S	U	U	D	D	D	D	D	D
5	D	S	U	D	D	D	D	D	D	D
6	D	S	U	U	U	D	S	U	U	D

uplink backhaul transmission into one GP if there is no much data to be transmitted on the backhaul link. So three multiplexing cases are possible:

1. Only downlink backhaul transmission multiplexed into special subframe.
2. Only uplink backhaul transmission multiplexed into special subframe;
3. Both downlink backhaul and uplink backhaul multiplexed into one special subframe.

Figure 4.25 shows the examples of these three cases [23].

As for general TDD systems and half-duplex relay, switching time has to be accommodated as well in Fig. 4.25. The switching time for eNB could reside in

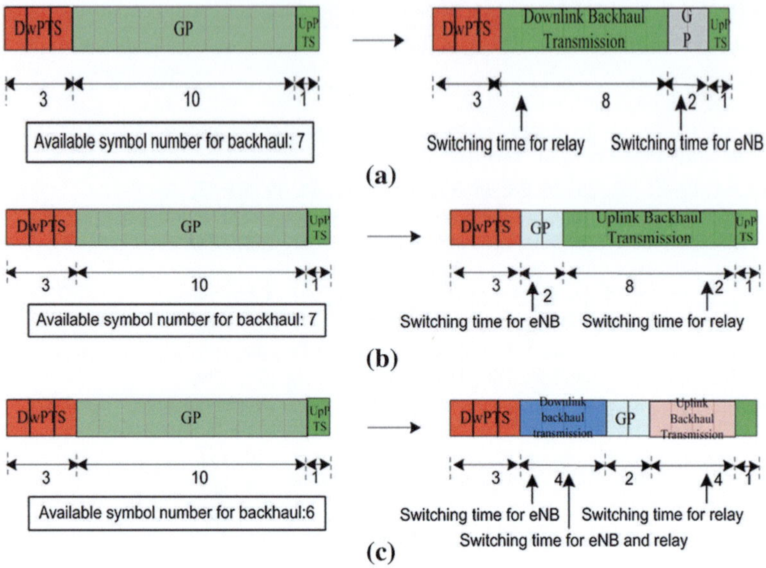

Fig. 4.25 Three cases of multiplexing DL and/or UL backhaul transmissions to a special subframe. **a** Only DL backhaul transmission in special subframe. **b** Only UL backhaul transmission in special subframe. **c** Both DL and UL backhaul transmissions in a special subframe

GP zone. For example, in Fig. 4.25a, eNB needs a Tx/Rx switching time between downlink backhaul transmission and UpPTS reception, and this switching time could resides in the first symbol of GP. At relay node, a Tx/Rx switching time is needed between DwPTS transmission and downlink backhaul reception, and the first symbol of downlink backhaul transmission may be used to ensure relay has enough time to do the switching. Thus for case 1, two symbols would be needed for Tx/Rx switching. Similar situation is seen in case 2 and case 3, except that one more switching time is needed in case 3. Therefore, the numbers of available symbols for backhaul transmission are 7, 7 and 6 in case 1, 2 and 3, respectively.

Using special subframe for backhaul transmission, however, has two major issues:

Issue 1: Limitation of backhaul propagation delay

As seen in case 1, relay needs to receive UpPTS from relay UEs after receiving downlink backhaul transmission, then the propagation delay between DeNB and relay should be less than 2 symbols as illustrate in Fig. 4.26. These 2 symbols for propagation delay would limit the DeNB-RN distance to about 20 km. The similar situation is seen in case 2, and the similar result could be obtained. In case 3, because relay need to start uplink backhaul transmission after receiving downlink backhaul transmission, then the round trip time between DeNB and relay is limited to 2 symbols. Then the coverage for case 3 is limited to $\sim$10 km.

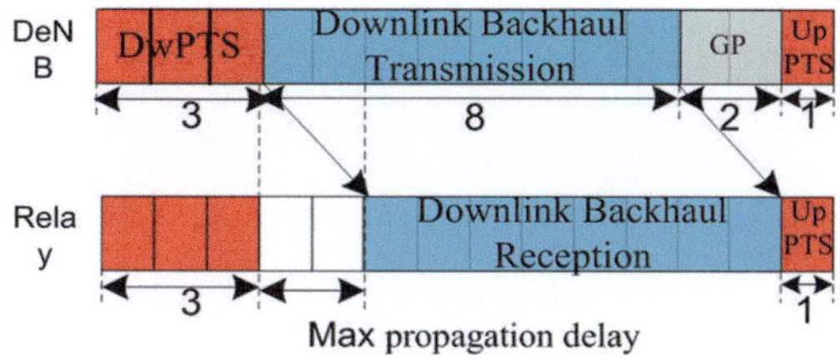

Fig. 4.26 Coverage issue for relay in case 1

Issue 2: Inter-site interference

Since part of the GP is utilized for backhaul transmission, if there is no coordination among eNBs about new special subframe pattern, the inter-site interference would among eNBs. For example if one eNB is configured as case 1 and another eNB is configured as case 2, then severe inter-site interference would occur in the overlapped part of downlink backhaul transmission and uplink backhaul transmission, as illustrated in Fig. 4.27. On the other hand, even there is no overlap, the transmit signal of one eNB could still leak to the other eNB receiving zone.

Consider the above issues of using special subframe for backhaul transmission, and the significant effort needed for the specification, RAN1 decided not to standardize this feature in Release 10 relay. Consequently, *TDD UL-DL configuration #0 and #5 are not supported in Release* 10 *relay.*

For the rest of UL-DL configurations, the backhaul subframe allocation follows the following principles.

- Support symmetric and downlink heavy allocations.
- Explicit allocation supported, to follow Release 8 principle for TDD subframe configuration.
  - Several Un subframe allocation patterns defined in each support TDD configuration, with different DL/UL subframe ratios to provide more deployment flexibility.
- RTT = 10 ms for supported configurations.
- Release 8 UL grant timing is reused, while Release 8 UL ACK/NACK timing is modified in some configurations. The motivation is to maintain an efficient scheduling at the donor eNB [24].
  - To minimize the modification of Release 8 UL ACK/NACK timing, i.e., with the least number of impacted subframes.
  - Release 8 ACK/NACK should occur at least 4 subframes after Un PDSCH transmission.

In UL-DL configuration 1, subframes #4 and #9 can be configured as DL backhaul subframes. Reusing Release 8 HARQ timings, we can allocate subframes #8 or/and #3 as UL backhaul subframes where both PDSCH to UL ACK/NACK timing, and PUSCH to UL grant timing can fit, as shown in Fig. 4.28. It corresponds to SubframeConfigurationTDD #4 in [3].

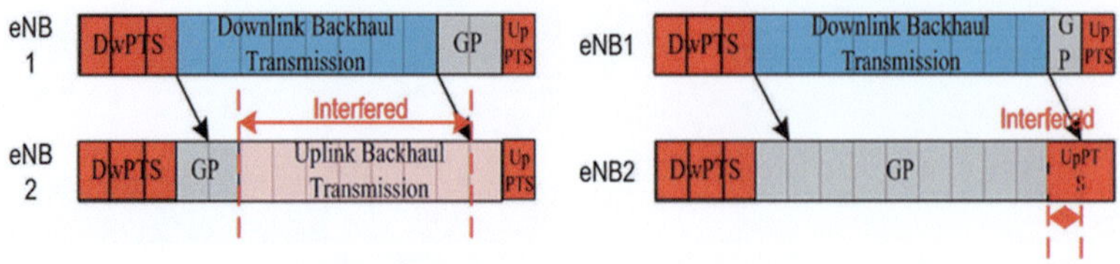

Fig. 4.27 An example of inter-site interference with different configurations of backhaul transmission in special subframe

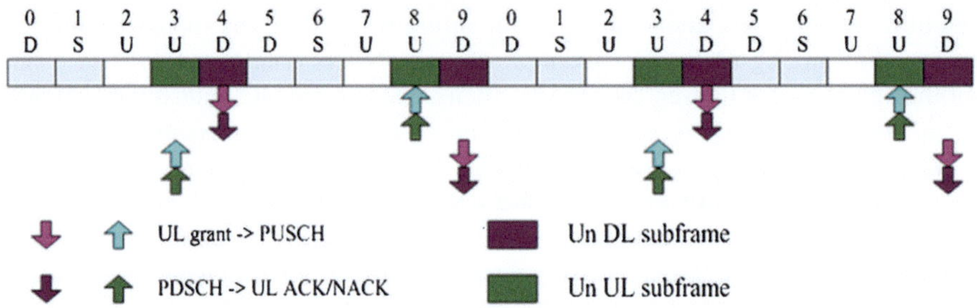

Fig. 4.28 An example of backhaul uplink HARQ timing for UL-DL configuration 1 (SubframeConfigurationTDD #4)

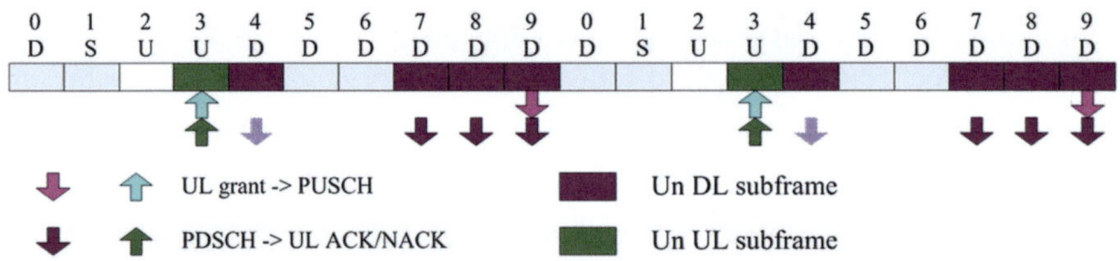

Fig. 4.29 An example of Un HARQ timings for UL-DL configuration 4 (SubframeConfiguration TDD#17)

Figure 4.29 shows an example of backhaul subframe allocation in UL-DL configuration 4. For UL-DL configuration 4, there are four DL subframes #4, #7, #8 and #9 can potentially be allocated as Un subframes, and subframes #2 and #3 can potentially be allocated as Un UL subframes. The example in Fig. 4.29 corresponds to SubframeConfigurationTDD #17 in [3] and can achieve DL:UL subframe ratio of 4:1. Here #4, #7, #8 and #9 can be configured as DL Un. From the prospect of UL grant timing, it does not matter whether subframe #2 or #3 is configured for UL Un. However, if subframe #2 is configured for UL Un, the HARQ timing relation between Un DL transmission in subframe #7, #8, #9 and Un UL ACK/NACK would not follow Release 8 timing. On the other hand, if subframe #3 is configured for UL Un, only UL ACK/NACK corresponding to DL Un

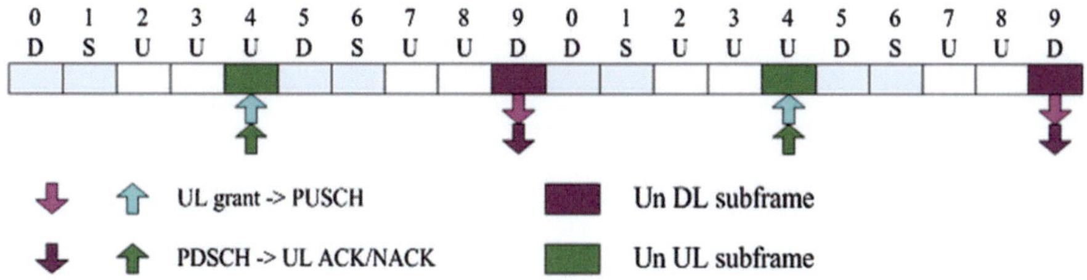

Fig. 4.30 An example of backhaul uplink HARQ timing for UL-DL configuration 6(Subframe-Configuration TDD#18)

Table 4.12 Backhaul subframe configuration table for TDD

Subframe ConfigurationTDD	eNB-RN uplink–downlink configuration	Subframe number *n*									
		0	1	2	3	4	5	6	7	8	9
0	1					D				U	
1					U						D
2						D				U	D
3					U	D					D
4					U	D				U	D
5	2			U						D	
6					D				U		
7				U		D				D	
8					D				U		D
9				U	D	D				D	
10					D				U	D	D
11	3				U				D		D
12					U				D	D	D
13	4				U						D
14					U				D		D
15					U					D	D
16					U				D	D	D
17					U	D			D	D	D
18	6					U					D

transmission in subframe #4 would not be able to reuse Release 8 HARQ timing. Therefore, only #3 is configured for UL Un to minimize the specification impact when DL:UL subframe ratio 4:1 is sought in UL-DL configuration 4. The modification of UL ACK/NACK is reflected in [3] where in Row #17, subframe #9 is added in the column "n = 3" compared to Table 10.1-1 in [25].

In TDD UL-DL configuration 6, the only available Un DL subframe is #9 as shown in Fig. 4.30. According to Release 8 HARQ timing, the corresponding PUSCH, as well as the UL ACK/NACK for PDSCH, is sent in SF #4. Hence, subframes {#4, #9} can be paired without violating Release 8 UL grant timing and UL ACK/NACK timing.

Table 4.13 Backhaul UL ACK/NACK timing in TDD

Subframe ConfigurationTDD	K according to subframe: $\{\kappa_0, \kappa_1, \ldots, \kappa_{M-1}\}$									
	$n = 0$	$n = 1$	$n = 2$	$n = 3$	$n = 4$	$n = 5$	$n = 6$	$n = 7$	$n = 8$	$n = 9$
0									4	
1				4						
2									4,9	
3				4,9						
4				4					4	
5			4							
6								4		
7			4,8							
8								4,8		
9			4,8,9							
10								4,8,9		
11				4,6						
12				4,5,6						
13				4						
14				4,6						
15				4,5						
16				4,5,6						
17				4,5,6,9						
18					5					

Table 4.14 Number of uplink HARQ processes for TDD

SubframeConfigurationTDD	Number of uplink HARQ processes
0–3, 5–18	1
4	2

However in Release 8, DL ACK/NACK associated with PUSCH in subframe #4 is sent in subframe #0, so the UL HARQ RTT is not 10 ms. Although R-PHICH will not be specified in Release 10, RN should know when to receive ACK/NACK (carried in UL grant) since UL HARQ in Un is still synchronous. Therefore, PUSCH to UL grant timing should be modified to match 10 ms RTT in the backhaul, as shown in Fig. 4.30.

For completeness, similar to the case of FDD, the number of HARQ processes for various TDD backhaul subframe configurations are also specified in [3], copied here as Table 4.14. Note that Table 4.14 can be solely derived from Table 4.12 and Table 4.13.

References

1. CMRI-NSN: Relay cost analysis performance and cost evaluation of relay deployment in CMCC Beijing networks, Nokia-Siemens Networks, CMCC, Dec 2010
2. 3GPP R1-094481: Necessity of R-PHICH, RAN1 #59, LG Electronics, Nov 2009
3. 3GPP TS 36.216: Evolved Universal Terrestrial Radio Access (E-UTRA): Physical layer for relaying operation
4. 3GPP R1-100977: Timing of backhaul and access link in downlink, RAN1 #60, ZTE, Feb 2010
5. 3GPP R1-100978: Timing of backhaul and access link in uplink, RAN1 #60, ZTE, Feb 2010
6. 3GPP R1-101774; Design of relay frame timing in LTE-A, RAN1 #60bis, CATT, CATR, CMCC, April 2010
7. 3GPP R1-100544: Synchronization in backhaul link, RAN1 #59bis, ZTE, April 2010
8. 3GPP R1-110214: Resource allocation schemes of R-PDCCH, RAN1 #63bis, Panasonic, Jan 2011
9. 3GPP R1-110809: Resource allocation schemes of R-PDCCH, RAN1 #64, ZTE, Jan 2011
10. 3GPP R1-106316: Un TBS determination, RAN1 #63, CMCC, CATT, Nov 2010
11. 3GPP R1-104651: Performance evaluation of Un DMRS, RAN1 #62, ZTE, Alcatel-Lucent, Alcatel-Lucent Shanghai Bell, Aug 2010
12. 3GPP R1-103834: R-PDCCH blind decoding for frequency diversity and frequency selective R-PDCCH transmissions, RAN1 #61bis, NEC Group, June 2010
13. 3GPP R1-103488: R-PDCCH design in LTE-A, RAN1 #61bis, CATT, June 2010
14. 3GPP R1-104339: Details on R-PDCCH interleaving in LTE-A, RAN1 #62, CATT, Aug 2010
15. 3GPP R1-105176: Design of non-interleaving R-PDCCH in Rel-10, RAN1 #62bis, CATT, Oct 2010
16. 3GPP R1-103492: Considerations on backhaul uplink transmissions, RAN1 #61bis, CATT, June 2010
17. 3GPP TS 36.212: Evolved Universal Terrestrial Radio Access (E-UTRA): Multiplexing and channel coding
18. 3GPP, R1-106163: Shortened format 2/2a/2b for Un PUCCH, RAN1 #63, Huawei, HiSilicon, Nov 2010
19. 3GPP R1-103121: Un HARQ timing for FDD, RAN1 #61, Huawei, May 2010
20. 3GPP R1-104858: Discussion on UL Un HARQ RTTs, RAN1 #62, Ericsson, ST-Ericsson, Aug 2010
21. 3GPP R1-104307: Un HARQ timing for uplink FDD, RAN1 #62, Huawei, Aug 2010
22. 3GPP TS 36.211: Evolved Universal Terrestrial Radio Access (E-UTRA): Physical channels and modulation
23. 3GPP R1-100342: Use of special time slot for backhaul link in TDD relay, RAN1 #59bis, Nokia, Nokia Siemens Networks, Jan 2010
24. 3GPP R1-104859: Further considerations on Un HARQ timing for TDD, RAN1 #62, Ericsson, ST-Ericsson, Aug 2010
25. 3GPP TS 36.213: Evolved Universal Terrestrial Radio Access (E-UTRA): Physical layer procedures

Chapter 5
Higher Layer Aspects and RAN4 Performance Aspects

5.1 Relay Architecture

Four alternatives were discussed for relay architecture [1] as listed below:

- Alternative 1:
  In this case, the relay node is a full L3 relay, transparent to the donor eNB. S1-AP is terminated at RN and data radio bearer (DRB) is used for transferring S1-AP. Release 8 header compression is reused on outer header only. QoS control can be at RN bearer granularity.
- Alternative 2:
  There is a S1/X2 proxy at the donor eNB, with additional Home eNB gateway functionality for relay evolved packet system (EPS) bearers. Release 8 header compression is reused on outer header only. QoS control is at RN bearer granularity.
- Alternative 3:
  RN bearers are terminated at the donor eNB, and S1-AP is terminated at RN. DRB is used for transferring S1-AP. Release 8 header compression is reused on outer header only. QoS control is at RN bearer granularity.
- Alternative 4:
  S1 U-plane terminated at the donor eNB. Signaling radio bearer (SRB) is used for transferring S1-AP. Release 8 header compression can be fully reused. QoS control is at UE bearer granularity.

There are a quite lot of discussions on pros. and cons. of those alternatives. The design trade-offs include the following aspects:

1. S1-AP/GTP-U termination point;
2. SRB or DRB to transfer S1-AP;
3. Simple control transmission protocol (SCTP)/IP on Un interface;
4. Mapping between UE EPS bearer and RN radio bearer

Y. Yuan, *LTE-Advanced Relay Technology and Standardization*, Signals and Communication Technology, DOI: 10.1007/978-3-642-29676-5_5,

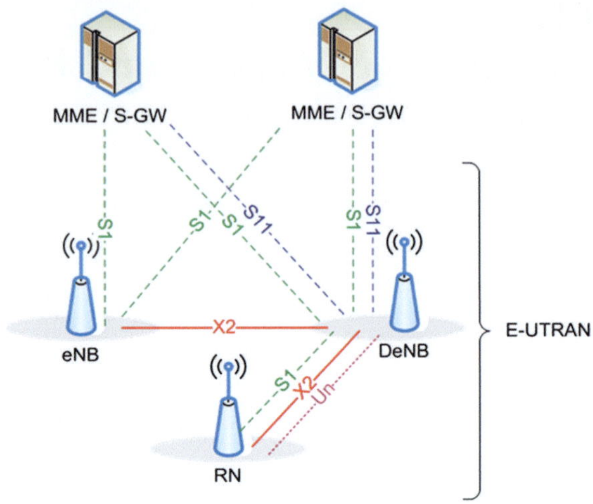

Fig. 5.1 Release 10 relay architecture

In Alternative 1 and Alternative 3, S1-AP and GTP-U are terminated only at the RN. There is no need to modify or forward S1-AP and GTP-U connections. In Alternative 2 and Alternative 4, S1-AP and GTP-U connections are also terminated at the donor eNB. That leaves room for optimizations. For example, Home eNB gateway functionality can be integrated in the donor eNB so that it can forward the packet during UE mobility. Such optimization, though quite desirable from performance point of view, would increase the design complexity and the development cost of the donor eNB.

Except Alternative 4, all other three alternatives favor using DRB to transfer S1-AP connections. Some concerns were raised regarding the security and latency of DRB which does not have integrity functionality. The scheduling priority of DRB is also lower than that of SRB. However, it was believed by a number of companies that integrity check functionality can easily be applied in DRB (if required) which is quite similar to SRB.

There was no consensus whether SCTP/IP is required on Un interference when SRB is used for transferring S1-AP signals. Some companies think that SRB can provide similar functions as SCTP, while others believe that the current design of SRB does not support the multi-streaming function required by the transport layer of S1-AP. Note that if SCTP/IP is used to transport S1-AP message on Un interface, an extra header compression scheme may be required to reduce header overheads which can be significant due to the excessive signaling in LTE network.

In Alternatives 1-3, EPS bearers of UEs connected to the same RN with similar QoS are mapped into the same RN radio bearer (many-to-one mapping). The justification is that for Un interface, static or semi-static QoS configuration/mapping should be sufficient in Release 10. In Alternative 4, each EPS bearer of UE connected to the RN is mapped to separate radio bearers over the Un interface (one-to-one mapping), thus providing UE bearer level granularity.

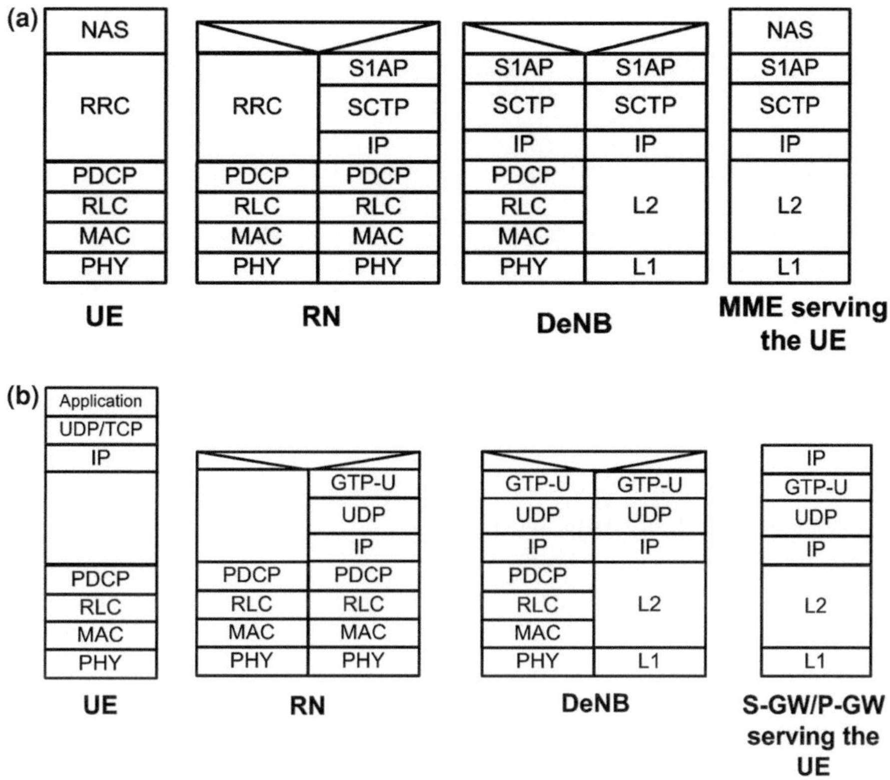

Fig. 5.2 **a** C-plane protocol stack of Release 10 relay. **b** U-plane protocol stack of Release 10 relay

Considering that (1) Release 10 relay nodes would primarily be deployed in fixed locations, (2) effective support of UE mobility is important, Alternative 2 was selected for Release 10 relay architecture which can be illustrated in Fig. 5.1.

Figure 5.2 shows the C-plane and U-plane architecture of Alternative 2. Here we use S1 interface as an example, and the principle applies to X2 interface as well. Here, S1-AP messages are terminated in both donor eNB and RN. The operation corresponds to an S1-AP proxy mechanism which is similar to home eNB gateway functionality. The S1-AP proxy operation is transparent to the MME and RN. To MME, it looks as if the UE is connected to the donor eNB. For RN, it looks as if the RN is talking to the MME directly. S1-AP messages encapsulated by simple control transmission protocol (SCTP)/IP are transferred over an EPS data bearer of the RN. In the U-plane protocol stack, GTP-U tunnels are terminated in both the donor eNB and RN. By integrating home eNB gateway like functionality, a donor eNB can perform S1/X2 proxy functions.

The bearer mapping is illustrated in Fig. 5.3. A radio bearer of RN can contain multiple UE radio bearers.

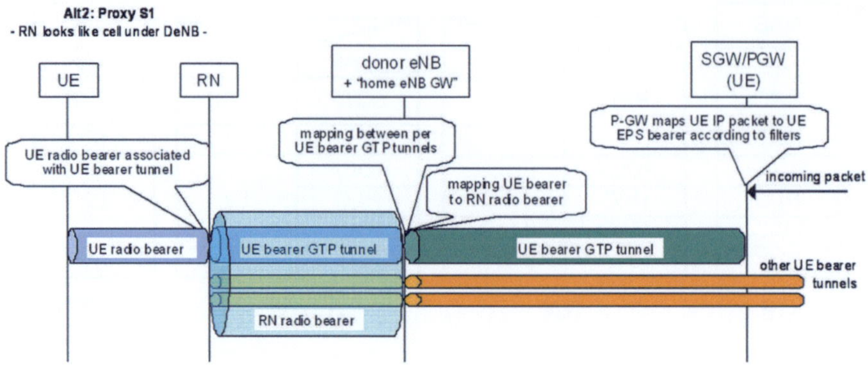

Fig. 5.3 Bearer mapping for Release 10 relay

Release 10 relay node RN is viewed as a cell (or sector) of the DeNB by LTE network and has the following functionalities:

- All eNB functions.
- Subset of UE functions. RRC is modified to support RN configuration. RACH remains the same, RN operation suspended while RACH is performed. NAS remains the same.
- Handles Uu inteface between UE and RN.
- Terminates S1/X2/Un interface. S1/X2 are mapped to RBs over Un. User plane data are mapped based on QCI of UE EPS bearer. The data mapping is configured by OAM and supports many-to-one mapping. Process non-UE-dedicated messages and UE-dedicated messages.

DeNB should support the following functionalities:

- Embedded S/PGW function for RN.
- Proxy for S1/X2 interface between RN and other network element (MME for S1 and other eNB for X2). Perform non-UE-dedicated messages.

5.2 C-Plane Procedures

There are two phases during RN startup. Phase I is to attach for RN preconfiguration, shown in Fig. 5.4. First, the relay node attaches to the E-UTRAN/ EPC as UE during power-up and retrieves initial configuration parameters, for example, the list of DeNB cells. Some initial configuration parameters can be obtained from RN OAM. The MME performs the S-GW and P-GW selection for the RN as a normal UE. RN obtains RN type information via pre-configuration or OAM. After this operation is complete, the relay node detaches from the network as a UE and triggers Phase II.

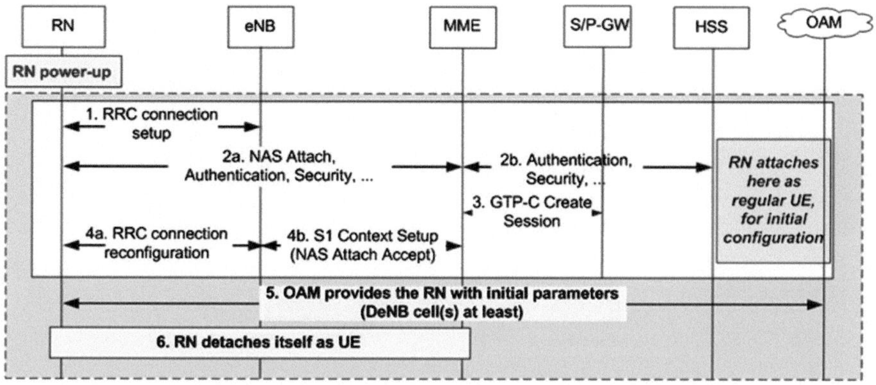

Fig. 5.4 Phase I of relay node startup

Some basic relay physical layer parameters can be configured via OAM, for example:

- Relay type:

  type 1 relay (inband relay, need to configure backhaul subframes)
  type 1a relay, outband relay, no need to configure backhaul subframes, needs to operate in two bands
  type 1b relay, can provide enough isolation between Un and Uu links, no need to configure backhaul subframes.

- The end symbol of DL backhaul subframe, UL relay frame timing, which depends on the synchronization requirement (FDD or TDD), and the propagation delay between DeNB and RN.
- The number of RN transmit antennas in Uu link. One OFDM symbol is needed for PDCCH in MBSFN subframes if this number is 1 or 2, otherwise two OFDM symbols are used for PDCCH. This information helps DeNB to configure the start symbol of DL backhaul subframes.
- The operating bandwidth of Uu link.
- The rating of the power amplifier in RN, especially for Uu link.

The operation in Phase II is to attach for RN operation, as shown in Fig. 5.5. The MME informs DeNB via S1 setup procedure, and RN indicates to DeNB via RRC connection establishment. RN is authenticated by MME. DeNB determines whether to configure Un subframes based on RN type, for example, Un subframes need to be configured for type 1 relay node which operates inband and has to share the time resources with Uu link. DeNB includes RN indicator and IP address of SGW/PGW integrated in Initial UE Message to MME.

The relay physical layer parameters to be configured by RRC messages are:

- Backhaul subframe configuration for TDD, with size of 5 bits and can take values from 0 to 18, see Tables 5.2 2 in [2].

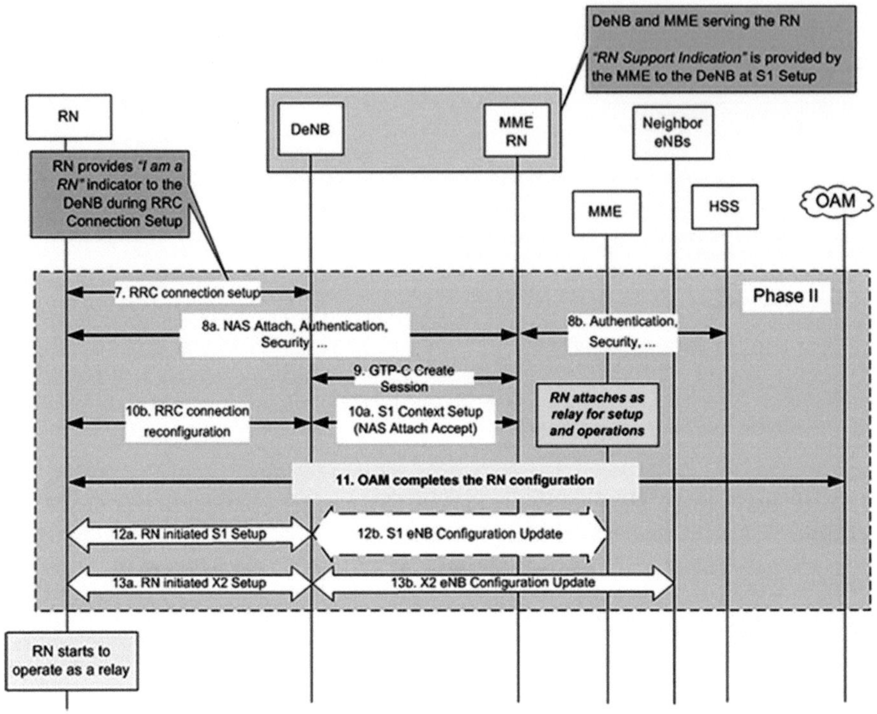

Fig. 5.5 Phase II of relay node startup

- Backhaul subframe configuration for FDD, is configured with an 8-bit bitmap, see Sect. 5.2 in [2].
- Start symbol index of DL backhaul subframes, with size of 2 bits and can take values 1, 2 or 3, see Table 5.4-1 in [2].
- Resource allocation type for configuration of the set of resource blocks for potential R-PDCCH transmission, with size of 2 bits and can take values 0, 1, or 3, see Sect. 5.6 in [2].
- Virtual to physical resource block mapping for configuration of the set of RBs for potential R-PDCCH transmission, with size of 1 bit and can choose between LVRB or DVRB, see Sect. 5.6 in [2] applicable to R-PDCCH resource allocation Type = 2 only.
- Resource block assignment for configuration of the set of resource blocks for potential R-PDCCH transmission, in the form of a bitmap equivalent to corresponding field in DCI.
- Interleaving mode of R-PDCCH, with size of 1 bit, either interleaving or no ininterleaving.
- Reference signal used for R-PDCCH demodulation, applicable to no interleaved R-PDCCH, with size of 1 bit, either common reference signal (CRS) or demodulation reference signal (DMRS).

Table 5.1 Comparison of two options of radio link failure handling for Un

	Option 1	Option 2 (supported)
O&M data	Lost and need to download again	Maintained
S1/X2 connection	Released	Can be kept
DRBs	Released	Can be kept
RN context and security parameters	Released	Can be kept or updated
UE connections	Released	Can be kept
Time to recover	Long	Short
Specification change	Little	Little

- PUCCH resource for ACK/NACK feedback on antenna port 0 and port 1, with size of 11 bits and can take values from 0 to 2047.

Since subframes #0 and #5 cannot be configured for Un subframes, RN should obtain the system information via dedicated RRC signaling. The system information in the RRC has complete SIB1 and SIB2, without MIB, nor SIB3–SIB9 (not to support mobile relay), nor SIB10-SIB12, nor SIB13 (MBMS not supported). There is no 3 h limitation for valid time of the system information.

Two options were proposed to handle Un radio link failure (RLF):

- Option 1: RN turns back to IDLE and performs initial startup procedure.
- Option 2: RRC connection reestablishment as RN.

Pros and cons of Option 1 and Option 2 are compared in Table 5.1, with the conclusion to support Option 2 only.

5.3 U-Plane Procedures

Two S1 bearers need to be set up. One is between S-GW and DeNB, the other is between DeNB and RN. Figure 5.6 shows the flow chart of the UE bearer setup. Since UE bearer is carried over Un DRB, the set up of UE bearer may lead to the modification of Un DRB.

The setup of RN radio bearer, or equivalently Un DRB which carries S1/X2 signaling, is considered as implementation. It can be performed in two ways as shown in Fig. 5.7. In Fig. 5.7a, P-GW/S-GW in DeNB initiates the set up via RN radio bearer after RN is attached. In Fig. 5.7b, RN requests the set up of RN radio bearer.

Release 10 relay nodes do not support MBMS. There is no flow control on Un. Header compression enhancements are not supported for Un in Release 10 specifications.

Semi-persistent scheduling on Un is not supported for type 1 RNs which requires a Un subframe configuration. The rationale is that the aggregate traffic of semi-persistent traffic of multiple UEs (served by RN) would no longer exhibit the semi-persistent nature.

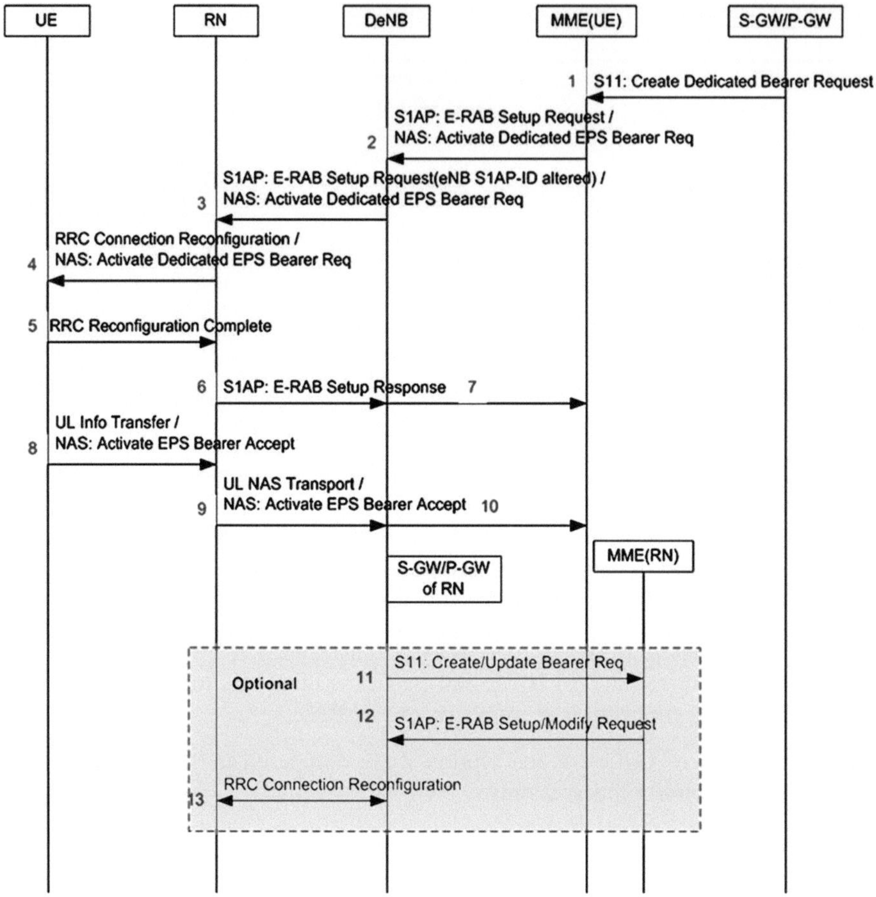

Fig. 5.6 Flow chart of UE bearer setup

5.4 S1/X2 Procedures

In S1 non UE associated procedure, if more than one RN is involved, the DeNB may wait and aggregate the response messages from all involved RNs before responding to the MME. In RESET case, the donor eNB does not need to wait the response message(s) from MME(s) or RN(s) before it responds with the RESET ACKNOWLEDGE message to the originating node. If S1 setup response involves one or multiple MME, RN would ignore those MME information. RN needs to know MME overload information to reject UE access.

In S1-UE associated procedure, S1-AP UE ID, TNL address and GTP TEID are to be changed by the donor eNB, for the purpose of UE associated S1 forwarding. Donor eNB forwards the handover resource allocation message to the target RN that is indexed by cell ID.

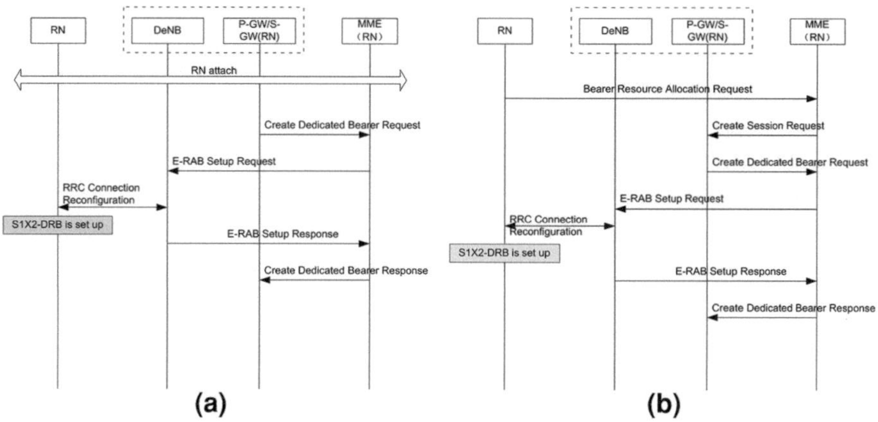

Fig. 5.7 **a** Flow chart for RN bearer setup via RN dedicated EPS bearer. **b** Flow chart for RN bearer setup via RN request

In X2 non UE associated procedure, it is up to donor eNB implementation whether the eNB configuration update is forwarded to RN.

In UE associated X2 forwarding, X2-AP UE ID, TNL address and GTP TEID are to be changed by donor eNB. If the handover source is an RN, the handover type is determined by the RN. X2 handover will be tried first. If it fails, S1 handover will be triggered. If the handover target is an RN, the handover type is determined by the sourcing eNB. Since the donor eNB of the target RN knows the target RN via ECG, it initiates the same handover type as the source.

5.5 Release 10 Relay Performance Aspects

Performance requirements for relay node includes:

- Co-existence study.
- Requirement for access link.
  RF requirements: ACLR, ACS, unwanted emission, reference sensitivity, blocking, spurious, intermodulation, etc.
  Baseband requirements: PUCCH, PUSCH, PRACH.
- Requirement for backhaul link,
  RF requirements: ACLR, ACS, unwanted emission, reference sensitivity, blocking, spurious, intermodulation, etc.
  Baseband requirements: R-PDCCH.

System simulation parameters for performance requirement study were largely borrowed from the performance study for RAN1, as discussed in Chap. 2.

Note that RAN4 work of relay was not completed in Release 10, e.g., only coexistence study was finished. The work was continued in Release 11 [3].

5.5.1 RF Requirements in General

For Release 10 relay RF performance, the principle for setting the requirements would be that:

- The access link performances are based on those of local area (pico) BS.
- The backhaul link performances are based on those of UE and local area (pico) BS, with necessary changes depending on the co-existence study. The reason for reusing eNB requirements is that Release 10 relay is usually deployed by the operators and does not move. Meanwhile relay access link has the similar output power/reference sensitivity with local area BS, lots of efforts could be leveraged from current specs.

Two power classes are defined for relay access link: maximum 24 dBm (Power class 1) and maximum 30 dBm (Power class 2) (Table 5.2). The rated power per antenna connector is scaled down with the number of transmit antennas, for example, in Power class 1, 24 dBm for single transmit antenna, 21 dBm for two transmit antennas and 18 dBm for four transmit antennas. For relay backhaul link, given the relatively good channel condition, only one power class is defined: maximum 24 dBm (Power class 1).

Related to relay power class, relay can be classified based on the RF scenarios for the access link deployment. Using minimum coupling loss criterion, the following relay classes can be defined:

- High coupling loss relay, characterized by requirements derived from outdoor relay scenarios with a relay to UE minimum coupling loss equal to 59 dB.
- Low coupling loss relay, characterized by requirements derived from indoor relay scenarios with a relay to UE minimum coupling loss equal to 45 dB.

RAN4 relay work was kicked off by coexistence study which focuses on the adjacent channel interference that is characterized by the adjacent channel leakage power ratio (ACLR) of the aggressor system and the adjacent channel selectivity (ACS) of the victim system. Together with ACS, the ACLR defines the adjacent channel interference ratio (ACIR) as

$$ACIR = \frac{1}{\frac{1}{ACLR} + \frac{1}{ACS}}$$

If aggressor and victim have different reference bandwidths, both ACLR and ACS definitions have to account for exactly those bandwidths.

The results of coexistence study are captured by the required ACIR for <5 % loss on average throughput and 5th percentile throughput. The following cases are found to be the most challenging:

Table 5.2 Relay node power classes

	Power class	Pmax (dBm)
Relay access link	Power class 1	24
	Power class 2	30
Relay backhaul link	Power class 1	24

Table 5.3 Backhaul antenna connector reference sensitivity

E-UTRA channel bandwidth (MHz)	Reference sensitivity power level (dBm)
1.4	−105.7
3	−102.7
5	−101
10	−98
15	−96.2
20	−95

1. The aggressors are the UE in uplink transmission and throughwall relay node in backhaul uplink transmission. "Throughwall" relay has one antenna outdoor for backhaul link communications and another antenna indoor for access link communications. The victim link is UE to eNB. The required ACIR should be higher than 50 dB.
2. The aggressors are the eNB and outdoor relay node in downlink access transmission. The victim link is eNB to UE transmission. The required ACIR should be higher than 50 dB.
3. The aggressor is eNB in downlink transmission. The victim links are eNB to outdoor relay node backhaul downlink transmission and eNB to UE downlink transmission. The required ACIR should be higher than 33.5 dB
4. The aggressor is the UE in uplink transmission. The victim links are UE to outdoor relay node access uplink transmission and UE to eNB uplink transmission. The required ACIR should be higher than 32.2 dB.

The conclusion of relay coexistence study is that ACLR requirement for relay backhaul transmitter shall be 45 dB. The operating band unwanted emission follows the requirement of local area base station for both category A and B specified in [4].

The carrier frequency error of each antenna connector relay backhaul should not exceed ±0.1 PPM observed over a period of one time slot (0.5 ms) compared to the carrier frequency received from eNB.

5.5.2 RF Requirements for Backhaul Link

For relay backhaul uplink transmission, the minimum requirement for error vector magnitude (EVM) is the same as the UE. That is: the average EVM level should not exceed 17.5 and 12.5 %, for QPSK/PBSK and 16 QAM, respectively. Similar to the case of UE, the requirement of EVM for 64 QAM is not specified.

Table 5.4 Access antenna connector reference sensitivity

E-UTRA channel bandwidth (MHz)	Reference measurement channel of Annex A.1 in TS 36.104	Reference sensitivity power level, PREFSENS (dBm)	
		PRAT = 24 dBm	PRAT = 30 dBm
1.4	FRC A1-1	−98.8	−98.8
3	FRC A1-2	−95.0	−95.0
5	FRC A1-3	−93.5	−93.5
10	FRC A1-3[a]	−93.5	−93.5
15	FRC A1-3[a]	−93.5	−93.5
20	FRC A1-3[a]	−93.5	−93.5

Note[a] $P_{REFSENS}$ is the power level of a single instance of the reference measurement channel. This requirement shall be met for each consecutive application of a single instance of FRC A1-3 mapped to disjoint frequency ranges with a width of 25 resource blocks each

The transmit intermodulation requirement defines the capability of the transmitter to inhibit the generation of signals in its nonlinear elements. For relay backhaul link transmitter, the requirement defined in Sect. 6.7 of [4] should be used.

The reference sensitivity of the backhaul link receiver is the same for all bands. Considering that a relay is typically connected to the power grid, the duplexer implementation can be better than for a UE terminal, so that lower noise factor can be achieved in the receiver chain. It is believed that approximately 1 dB performance improvement is feasible. The reference sensitivity is listed in Table 5.3.

5.5.3 RF Requirements for Access Link

The error vector magnitude (EVM) requirement for the access link follows the requirement for eNB which is 17.7, 12.5 and 8 % for QPSK, 16QAM and 64QAM, respectively.

The reference sensitivity requirement for the access link is the same for all the operating bands and listed in Table 5.4.

The relay access link blocking level reuses that of local area eNB.

5.5.4 Baseband Requirements

For R-PDCCH, formats with and without cross-interleaving, resource mapping with LVRB and DVRB, normal and reduced DMRS, are all supported in Rel-10. The performance requirements for all these R-PDCCH cases would be defined based on DL grant. The performance requirements for R-PDCCH with no cross-interleaving have been prioritized, with the focus on LVRB resource mapping and DMRS for demodulation.

Table 5.5 Test parameters for single-layer R-PDCCH transmission on antenna port 7

Parameter		Unit	Test 1	Test 2
Cyclic prefix			Normal	
Cell ID			0	
Un subframe type in DeNB			Normal subframe	
Subframe configuration			10110101 (FDD) 4 (TDD)	
Number of OFDM symbols for PDCCH		OFDM symbols	2	
Configuration of OFDM symbols for eNB-to-RN transmission in the first slot			2 (Note 1)	
DL power allocation	R-PDCCH_RA	dB	0	
	R-PDCCH_RB	dB	0	
Cell-specific reference symbols			Antenna port 0	Antenna port 0,1
CSI reference signal configuration			1	
Number of CSI reference signals configured			1	4
CSI reference signal subframe configuration			$I_{\text{CSI-RS}} = 37$	
N_{oc} at antenna port		dBm/15 kHz	−98	
Number of allocated resource blocks		PRB	2	4
Symbols for unused PRBs			OCNG (Note 2)	
Simultaneous transmission (Note 3)			No	
Beamforming model			No precoding	A precoder vector $W(i)$ of size 4×1 is randomly selected with the number of layers $\upsilon = 1$ from Table 6.3.4.2.3-2 in TS 36.211 as beam forming weights
Precoder update granularity			Frequency domain: 1 PRG Time domain: 1 ms	

Note 1 as specified in Table 5.4 in TS 36.216

Note 2 These physical resource blocks are assigned to an arbitrary number of virtual UEs with one PDSCH per virtual UE; the data transmitted over the OCNG PDSCHs shall be uncorrelated pseudo random data, which is QPSK modulated

Note 3 The modulation symbols of the signal under test are mapped onto antenna port 7 while antenna port 8 is unused

Note 4 $n_{\text{SCID}} = 0$

Table 5.6 Minimum performance for R-PDCCH without cross-interleaving (FRC)

Test #	BW (MHz)	Aggregation level	DCI format	Propagation Condition	Antenna configuration	Reference value	
						Pm-dsg (%)	SNR (dB)
1	10	2 PRB	Format 2C	LOS with strong dominant component	1 × 2	1	2.1
2	10	4 PRB	Format 2C	NLOS with medium correlation	4 × 2	1	11.5

The test parameters for single-layer R-PDCCH transmission on antenna port 7 are listed in Table 5.5.

As Table 5.5 shows, for FDD, the backhaul subframe configuration bitmap is 10110101. For TDD, the backhaul subframe configuration is the configuration #4, i.e., SF #4 and #9 for backhaul downlink, and SF #3 and #8 for backhaul uplink. The minimum performance for R-PDCCH without cross-interleaving is listed in Table 5.6.

5.5.5 Synchronization Requirements

The cell phase synchronization accuracy for a relay is defined as the maximum absolute deviation in frame start timing between the relay's access link DL transmission and its donor cell's DL transmission. A relay may support one of two synchronization cases: DL Case 1 and DL Case 3 as mentioned above. The minimum requirement for DL Case 1 is (17 μs + T_{prop}) where T_{prop} is the propagation delay between relay and its donor cell, regardless of cell sizes. For DL Case 3, the synchronization requirement is 3 and 10 μs for small cell (i.e., <=3 km) and large cell (i.e., >3 km), respectively.

References

1. 3GPP TR 36.806: Evolved Universal Terrestrial Radio Access (E-UTRA): Relay architectures for E-UTRA (LTE-Advanced)
2. 3GPP TS 36.216: Evolved Universal Terrestrial Radio Access (E-UTRA): Physical layer for relaying operation
3. 3GPP TR 36.826: Evolved Universal Terrestrial Radio Access (E-UTRA): Relay radio transmission and reception (release 10)
4. 3GPP TS36.104: Base station (BS) radio transmission and reception

Chapter 6
Implementation Aspects of Release 10 Relay

Relay implementation work includes the following aspects:

- Building a relay node that is capable of backhaul communications with donor eNB, and serving Release 8 UEs within its coverage
- Implementing appropriate radio resource scheduling and management at donor eNB to efficiently manage the backhaul resources for relay nodes, and possibly the marco UEs at the same time
- Relay node site planning, relay network optimization, budget analysis, etc.

We in this chapter will discuss those implementation aspects. A key assumption is that Release 10 relay is basically a module of network equipment. Therefore, its development is similar to building an eNB, or pico cell, rather than customer premise equipment such as femto node. Relay node implementation would be based on the framework of an eNB, with necessary changes to fulfill the functionalities of two hop decode-and-forward communications. The changes are mostly in the wireless backhaul link.

6.1 General Consideration of PHY Layer Implementation

A relay node communicates with donor eNB via backhaul (Un) link and with UEs via access (Uu) link. Thus it has both eNB and UE functionalities as Fig. 6.1 shows. Inband relay of FDD is assumed here. As discussed in earlier chapters, while an out-band relay can almost reuse all the baseband implementation of eNB and UE, it generally requires a separate radio for the out-band communication with DeNB, a significant additional cost of RN hardware. It is also at the expense for operators' spectrum. Note that the RF duplex shown in Fig. 6.1 would not be needed in TDD where DL and UL operates in the same band.

Y. Yuan, *LTE-Advanced Relay Technology and Standardization*,
Signals and Communication Technology, DOI: 10.1007/978-3-642-29676-5_6,

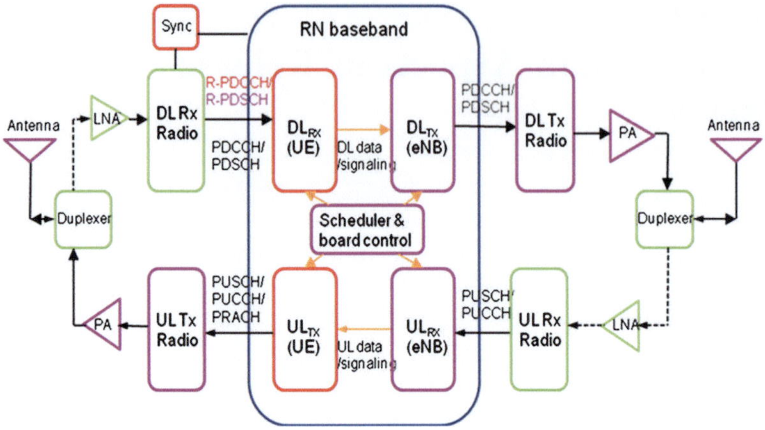

Fig. 6.1 eNB and UE functionalities in a Release 10 relay (inband FDD)

Figure 6.1 is a very high-level layout of the baseband, radio card and antennas. The red color indicates major changes needed for those modules compared to Release 8 eNB implementation or they simply do not exist in eNB. Pink (or magenta) color means that some modifications may be needed to make the relay node to be cost competitive and easy to deploy. Such change may also reduce the size and weight of a relay node so that it can be more easily deployed, e.g., on lamp posts. Light green color is used in those blocks where very limited changes are expected.

Note that a Release 10 relay node would behave like a Release 8 UE when the relay starts up or there is a radio link failure in Un link. Thus, a relay node should have basic functions of Release 8 such as broadcast channel (BCH), synchronization channel (SCH), CRS based demodulation, transmit diversity, etc. so that it can perform the cell search and radio link setup.

We will discuss in the following the impact on the implementation for each modules. To support relay, some modification is also needed in the donor eNB implementation, which will be addressed.

6.2 Baseband Realization of Relay Node

In the baseband, UE functionality needs to be added: DL reception and UL transmission. Since Release 8 PUCCH and PUSCH are reused in backhaul uplink, in principle the baseband implementation of Release 8 UEs may serve a reference. However, there are some practical issues:

- eNB vendors usually do not have detailed knowledge of baseband implementation UE, for example, channel estimation, channel decoding, signal to noise ratio estimation, internal timing, etc.
- UEs often have their own implementations in baseband, which would be very different that of eNB.
- Cost structure is different between eNB and UE. eNB is supposed to have more processing power and ought to serve multiple users. So the cost of less of concern. On the other hand, UE is a single user device. It has to be cheap and power efficient. Cheap price comes from economic scale–mass production, meaning that the UE chipsets are usually in ASIC. Compared to FGPA or DSP, longer design period and more development effort are needed for ASIC implementation.
- Compared to UEs, less number of RNs is expected to be deployed per donor eNB, meaning that there is less motivation of ASIC development.

From implementation point of view, the baseband development work at the transmitter is relatively easier: just to follow the specification and less need for proprietary algorithms. Therefore, a more practical way would be to use the off-the-shelf standard modules of FPGA, or DSP for UL backhaul transmission and DL access (Uu) link transmission.

Among the four DL/UL baseband Tx/Rx modules in the relay node, DL receiver module may need the most effort of development as R-PDCCH is a brand new physical channel introduced in Release 10. Also, the backhaul channel may need to support some of or all the transmission modes in Release 10, depending on the relay node capabilities. The more the transmission modes are to be supported in the backhaul, the more complex the development of DL receiver module. Therefore, FPGA or DSP implementation can be a starting point for some general algorithms of channel estimation, channel decoding, signal to noise ratio estimation that are applicable to both eNB and UE, and then followed by finer tuning and optimization of the implementation.

UL transmission module may need some work of development in certain RN capabilities to support Release 10 transmission mode for uplink single user MIMO. However, as mentioned above, transmitter side baseband development in general is a standard process, and requires less effort than for a receiver development.

For backward compatibility, Release 8 specification is reused in access link. So for eNB functionalities in RN such as DL transmission and UL reception Release 8 eNB implementation can readily be reused. However, since Release 10 relay is primarily used for coverage extension, it is expected that the number of UEs served by a RN would be smaller than the number of UEs served by an eNB. So in order to build a cost effective relay node to better match the typical deployment scenarios, the corresponding chipsets may need to be dimensioned differently from that for eNB. Again, FPGA or DSP seems more suitable for the chipsets.

In the following we discuss in more details of a few special considerations for the baseband implementations Release 10 relay node.

6.2.1 Channel Characteristics of Backhaul and Access Links

As discussed extensively in Chap. 2, relay node is typically placed on a lamp-post, or some constructions with media heights of 5–8 m. Relay node is a piece of operator-deployed network equipment and its site can be optimized to ensure better propagation environment, such as dominant line-of-sight components, strong signal strength, less interference, slow Doppler, etc. The large scale fading models in [1] can serve as a standard tool for quantitative evaluations of backhaul channel, especially for system performance evaluation. While there is no agreement on the fast fading model for backhaul performance study, in particular for the link level simulations, the models proposed in [2] provide the guidance on how to characterize the small scale fading in the backhaul link. They capture not only the delay spread information which determines the frequency selectivity of the channel, but also the spatial information such as angle spread which is crucial for the study on MIMO technologies. The cross-polarization parameter for example, XPD, helps to evaluate the antenna diversity in cross-polarizations.

Compared to macro-UE channel, the backhaul channel has less delay spread and the frequency response is in general less rigged. In another word, there are more correlations between channels in adjacent PRBs. Due to the fixed deployment of relay node, the Doppler is rather small in the backhaul. So the time correlation of the channels between adjacent subframes, or between adjacent OFDM symbols is high. Hence, the baseband reception development should take into account the time and frequency characteristics of the backhaul channel, so that the algorithms fit the channel.

Strong LOS component in backhaul propagation environment seems to favor the beamforming technologies. Beamforming antennas are usually closely-spaced and therefore require less space for installation, which is good for lamp-post mounted relay nodes. The baseband development may prioritize beamforming technologies, if antenna calibrations can be made cheap and effective. The high rank MIMO transmission, although with higher peak rate, may find less usage in if the backhaul propagation is scatterer-deficit. Rank of 3 and 4 transmission requires far-apart antennas at least at the relay node. Depending on the angle spread statistics of the channel, the requirement for "far-apart" may be 4 to 10 of wavelength, or even larger, translated in 0.6–1.5 m or beyond, if operated in 2 GHz. That would put significant constraint on the relay node deployment. So the development of high rank MIMO mode could be put in the second priority.

Compared to eNB to UE links, the access link would experience more severe pathloss compared to direct link, since the RN antenna height is much lower than that of base stations which is typically 25–30 m. Also as discussed in Chap. 2, fast fading of access link has its own characteristics, in terms of delay spread, angle spread and cross-polarization XPD. These channel characteristics can be taken into account during the development of resource scheduler in relay nodes.

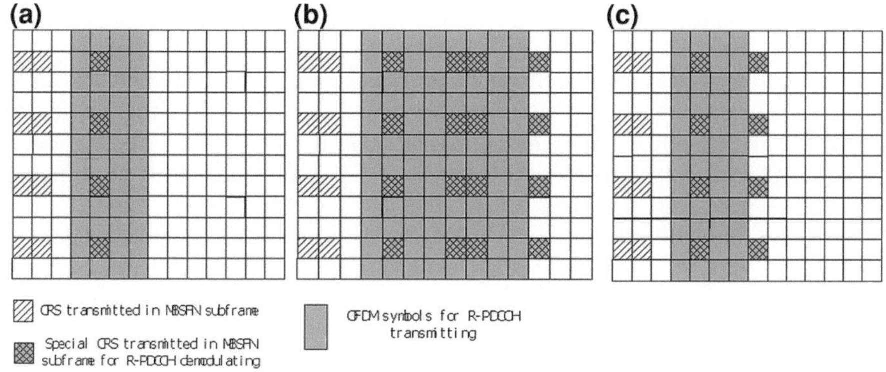

Fig. 6.2 Examples of CRS used for channel estimation of R-PDCCH in the first slot

6.2.2 Common Reference Signal Demodulation

CRS serves both the demodulation and channel measurement purposes, and therefore spans the entire operating band and a subframe. Such evenness distribution helps the channel estimation and eases the receiver implementation. However, for the backhaul link, the first one or two OFDM symbols cannot be received by the relay node since it needs to transmit PDCCH to the relay served UEs at that time, and also needs some time to switch from transmission to receive. The first one and two OFDM symbols contain CRS which cannot be used by RN for channel estimation.

For R-PDCCH, the DL grants are placed in 1st slot only in order to reduce the decoding latency. Depending on the actual implementation, partial or full CRS (except those in the first or two OFDM symbols) may be used for channel estimation as Fig. 6.2 shows. In the left figure (a), CRS only in the fifth OFDM symbol are used. It has the least decoding latency as the decoding of the DL grant(s) can start right after receiving the first slot. The drawback is that the limited number of CRS in time domain may degrade the channel estimation accuracy, either when interference/noise is strong, or when there is certain Doppler in the channel fading. In the middle figure (b), CRS in the fifth, eighth, ninth and twelfth OFDM symbols can be used for channel estimation. It leads to the longest decoding latency as the decoding of DL grant can only start after RN receives the whole subframe. The channel estimation is the most robust in (b). In the right figure (c), CRS in the fifth and eighth OFDM symbols are used for channel estimation. It is a compromise between the decoding latency and channel estimation accuracy. No matter which implementation is used for CRS demodulation, the channel estimation algorithms need to be modified or developed for the backhaul DL reception.

Besides for demodulation, CRS is also crucial for various measurements such as RSRP, CQI and PMI. CRS is present in all the subframes except those configured as MBSFN subframes. So for Release 8/9 UEs, CRS is continuously transmitted with few gaps. Such behavior can be exploited by UE receiver implementations where time domain smoothing and averaging are carried out to refine the channel estimation. But Release 10 relay node works in half-duplex mode and the backhaul subframes are not continuous in time. The backhaul subframe allocation patterns are quite diverse in FDD, there can be 1–6 subframes per radio frame for the backhaul.

In this sense, the backhaul link bears some similarity to TDD systems where DL subframes and UL subframes can sit side-by-side. TDD also supports a large number of DL/UL subframe ratios. Hence, channel estimation schemes for TDD UEs can be a good reference.

As discussed in Chap. 4, in the case that GPS synchronization is not possible, and it is not feasible to offset the subframe indices of backhaul link and access link, synchronization of relay node to donor eNB may be through CRS. Proprietary algorithms may be needed to achieve good accuracy of synchronization. In TDD, CRS almost has to be used in the absence of GPS, since subframe offset would introduce severe site-to-site interference. Given the rather stringent synchronization requirement of 3 μs and relatively sparse CRS in a subframe, it would be challenging job.

6.2.3 DL DMRS Demodulation

Different from CRS, the purpose of DMRS is solely for channel estimation for data channel demodulation. DMRS is only present in resources that carry the data channel and R-PDCCH if it is configured for DMRS. So in principle, algorithms for DMRS would be different from those for CRS.

For relay backhaul, the situation is a little more complicated as there are two DL DMRS patterns: (1) Release 9/10 DMRS present in both slots in DL timing case 1; (2) Release 9/10 DMRS present only in the first slot in DL timing case 3. Half of DMRS cannot be used in DL timing case 3 and it has some impact on the performance as shown in Chap. 4. Therefore, certain optimization may be needed in the baseband implementation so that the channel estimation accuracy can be maintained as much as possible comparable to DL timing case 1.

When DMRS in both slots can be received, it is still up to the implementation whether only DMRS in the first slot are used for DL grant demodulation. There is a trade-off between the channel estimation accuracy and decoding latency, similar to the discussion of CRS demodulation.

Since DMRS is UE or RN specific, and present only in PRBs that carries the data and R-PDCCH for that UE or RN, the average window in frequency domain in general should not go beyond each PRBs. Compared to CRS which spans over entire frequencies, DMRS has the drawback in not being able to average the channel estimates over adjacent PRBs in order to improve the estimation accuracy. It is not possible in Release 8 and 9 since the receiver cannot make sure that the

PRBs to carry its PDSCH are consecutive in frequency domain. In Release 10, PRB bundling is introduced to notify whether PDSCH spans continuously over several PRBs so that channel estimation can be refined in DMRS. That PRB bundling feature can be used for the backhaul link also. So the development of DL reception module at RN should support that feature.

6.2.4 Search Space for R-PDCCH Without Cross-Interleaving

For non cross-interleaved R-PDCCH, the same set of resource blocks is configured for the potential R-PDCCH transmission in the first and the second slots. This is different from cross-interleaved R-PDCCH where the resource blocks for DL grants in the first slots and the UL grants in the second slots may be different. The rationale of using the same resource blocks for non cross-interleaved R-PDCCH is to fully utilize the frequency selectivity and precoding gains which are supposed to apply both slots in a PRB pair. Putting them in slots of different PRB pairs would generally defeat the above purpose. The situation is a different in cross-interleaved R-PDCCH for two reasons: (1) diversity is the first priority which does not require the same PRB pairs for both DL grants and UL grants; (2) cross-interleaving allows more efficient packing of DL grants and UL grants. Due to less number of OFDM symbols in the first slot, i.e., 4, compared to that in the second slot, i.e., 6–7, and DL grant often carries more payload bits than UL grant, more PRBs is needed in the first slot to accommodate DL grants than UL grants in the second slot. Hence, two independent searches are needed for DL grants and UL grants decoding in cross-interleaved mode.

For R-PDCCH without cross-interleaving, the actual number of blind decodes can be made smaller by taking into account of the resource allocation characteristics. Essentially, if a PRB pair contains both DL grant in the first slot and UL grant in the second slot, those grants should target for the same relay node. So effectively, a relay node configured in non cross-interleaved mode may try to decode the data in the second slot of a PRB pair where its DL grant is already detected in the first slot, in order to search for the potential UL grant for it. By doing so, the number of blind decodes for the UL grants can be reduced to the minimal, or even deterministic.

The real situation may be more complicated if the resources used in the first slot and the second slot are highly imbalanced, for example, the DL grant carries many bits, but the second slot has 7 OFDM symbols, resulting in for example, 2–3 PRBs in the first slot, vs. 1 PRB in the second slot. In that situation, more blind decodes would be needed.

Although it is quite likely that the UL grant would be in the same PRB pair as DL grant, the specification does not mandate it. There might be situations that UL grant and DL grant are deliberately transmitted in different PRB pairs as Fig. 6.3 even if the R-PDCCH is not cross-interleaved. There is certain resource waste in the first slot if it is not used for transmitting DL grant for the same RN. However,

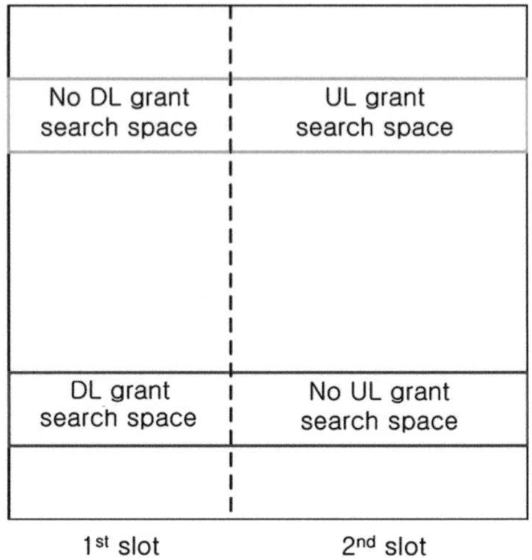

Fig. 6.3 An example of using different PRB pairs for DL and UL grants of R-PDCCH without cross-interleaving

the gain by assigning PDSCH in the second slot of a PRB that has DL grant in the first slot may outweigh the resource loss for R-PDCCH. To handle this situation, a relay node should not stop decoding UL grant if it does not detect its UL grant in the PRB pair that contains its DL grant.

6.2.5 Choice for Relay Timing

In Release 10 relay, there is no specification for uplink timing, since Release 8 already supports the shortened uplink subframe in which the last SC-FDMA symbol is punctured for SRS transmission. Relay uplink timing would end up with either the full use of 14 SC-FDMA symbols, or the shortened subframe in UL backhaul. It is up to the implementation on when and how to achieve that. Since there is no uplink timing specified, the standard would not mandate which downlink timing and uplink timing schemes should be paired, even though some choices are intuitive.

For FDD systems, there are two UL timing choices to pair with DL timing case 1: UL timing case 2b or UL timing case 3. In UL timing case 2b, all 14 SC-FDMA symbols can be used for UL Un. It has the benefit of maximizing the backhaul UL capacity, which is further ensured with less frequent SRS transmission due to the relatively stable RF condition of relay backhaul. However, the RN needs to configure the Uu subframe preceding the Un UL subframe to be cell specific SRS subframe. This would effectively reduce the occurrence of the SRS that are actually transmitted, thus negatively impact the RN cell UL capacity, in particular when there are a large number of UEs served by the RN. So UL timing case 2b is more suitable for stable backhaul and each RN has small number of UEs that are stationary or slow moving. On the other hand, Un subframe has to be configured as

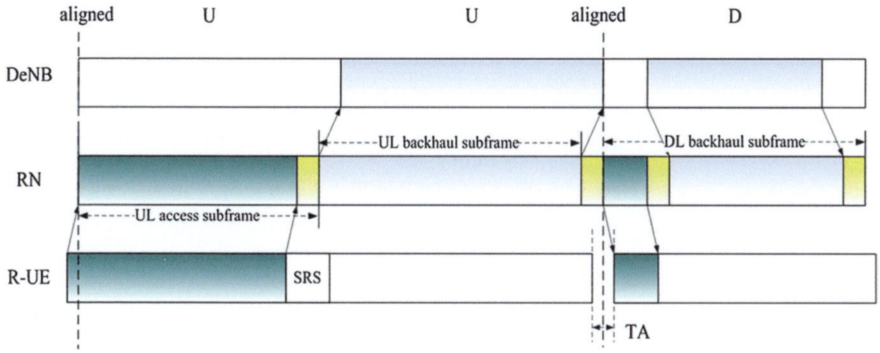

Fig. 6.4 DL timing case 3 and modified UL timing case 4 in TDD

SRS in UL timing case 3, which would reduce the backhaul UL capacity by 7 %. But the complete Uu UL subframe would benefit the uplink channel measurement with SRS. So UL timing case 3 is more suitable for the case when RN backhaul is not the limiting link and there are many moving UEs served by a RN.

For TDD systems, a natural choice of DL and UL relay timing is to pair DL timing case 3 and UL timing modified case 4. Since relay access link transmit timing and reception timing are exactly aligned, with respective to donor eNB transmit and reception timing as Fig. 6.4 shows, the effective length of GP would be maintained, i.e., the coverage of RN cell would not be reduced.

As discussed earlier, due to different relay timing cases, the number of symbols available in the backhaul subframe are different and DMRS patterns are not the same. So ideally, the implementation can be tailored for each situation, for example when DMRS of both slots or only in the first slot are available. On the other hand, common design flexible enough to fit most typical situations is also a reasonable choice from the reducing the development cost and product testing point of view.

The baseband implementation impacts are summarized in Table 6.1.

6.3 Radio Modules and Antennas of Relay Node

Besides the baseband, RF modules and antennas contribute the major development and hardware cost of a relay node.

6.3.1 Power Amplifier and Filters

Major change is seen in the transmitter. For the access link, the maximum transmit power of RN is typically about 23–30 dBm in 10 MHz, which is 16–23 dB lower

Table 6.1 Summary of baseband implementation impacts for Release 10 relay

Baseband module	Implementation
DL reception	New item, FPGA or DSP, general algorithms for UE receivers, no specific reference design as R-PDCCH is a new channel introduced in Rel-10, to take into account backhaul channel fading, RN timing, etc
UL transmission	New item, FPGA or DSP, follow Rel-8 specifications for UE transmitter
DL transmission	Can reuse Release 8 eNB transmission, possible with different dimensioning, less number of UEs compared to in eNB cell. Possible of using FPGA or DSP
UL reception	Can reuse Release 8 eNB reception, possible with different dimensioning, less number of UEs compared to in eNB cell. Possible of using FPGA or DSP

than that of eNB. Lower power amplifier (PA) can be used that is cheaper, less bulky, and consumes less power. Less distortion is expected with small PA class, meaning that

- RF filters can also be cheaper and light weighted with more relaxed requirement for spurious emission.
- Special circuit in radio board such as amplifier pre-distortion may be turned off if the frequency emission masks can be met.

For the backhaul link, the maximum transmit power should be comparable or even lower than that of UE, e.g., 23 dBm, compared to that of eNB (46 dBm). Therefore, the RF circuit designed for eNB transmitter can be further relaxed such as

- Use general purpose RF filters of low cost, small size and light weight.
- Use very low power amplifier class.
- Turn off pre-distortion circuit on RF board.

The RF receiver sees less impact. Most circuits can be reused from Release 8 eNB implementation, especially access uplink reception. For the backhaul downlink reception, the requirement for low noise amplifier (LNA) may be relaxed, considering the much higher transmitting power of eNB than UE, and better channel quality of the backhaul link.

The design experience from TDD can be utilized to address any potential issues with regard to Tx/Rx and Rx/Tx switching in RF circuits to support half-duplex operation at relay node.

6.3.2 Clock Synchronization

The initial synchronization is done during the startup stage when the RN operates as Release 8 UE to look for synchronization channel and etc. Once the RN starts to serve its UEs, it would use CRS in backhaul subframes to track the donor eNB timing if GPS is not available or subframe index offset is not allowed. The synchronization circuit (via air interface) is a new item compared to Release 8 eNB implementation. The circuit itself would involve both baseband board and RF modules.

Note that through LTE air interface we can only achieve relative synchronization to donor eNB, e.g., with propagation delay in the backhaul link. For absolute synchronization, such propagation delay has to be subtracted out via certain means, for example, timing-advance information in UL. Alternatively, Global positioning system (GPS) could be used that can provide accurate global synchronization albeit requiring another set of circuit and may not work indoor, or in tunnels.

6.3.3 Antennas

As discussed in Chap. 2, the typical antenna of relay node on backhaul link is directional e.g., 70°, of 7 dBi gain. The horizontal beam-width is comparable to that of macro eNB, therefore, the antenna backplane design for Release 8 eNB can be reused here. For the access link, horizontal omni-directional antenna is often used with 5 dBi gain.

The much lower gain of RN antennas compared to eNB antennas (7 vs. 17 dBi) implies that RN antennas have much wider beam-width vertically. In another word, the antenna size (in height) can be much shorter than that of eNB antennas, with less number of vertically spaced antenna elements. The fatter beam in vertical direction helps to reduce the size of antennas so that the relay node can be more flexibly deployed. Also it makes down-tilt exercise easier. Imaging the relatively low installation of outdoor relay, for example, 5–8 m high, with 15–20° beam-width in vertical, there is not much difference whether the antenna down tilted by 10 or 20°.

Depending on the carrier frequency and designed coverage area, antennas on wireless routers for consumers may be used here to allow easy and low-cost deployment.

The RF and antenna implementation impact for Release 10 is summarized in Table 6.2.

6.4 Relay Node Scheduler

Relay node has to support Release 8 UEs and therefore much of the implementation for Release 8 eNB scheduler can be reused. However, there are a few exceptions to be considered so that the relay node can be effective in coverage extension and even possibly improve the system capacity.

Table 6.2 Summary of RF and antenna implementation impact

RF & antenna module	Implementation
DL transmitter	Reduced version of Rel-8 eNB transmitter, with cheaper RF filters, low power amplifier rating, possible no pre-distortion
UL transmitter	Much more reduced version of Rel-8 eNB transmitter, very cheap RF filters, very low power amplifier rating, no pre-distortion
UL receiver	Can reuse Rel-8 eNB receiver implementation
DL receiver	Can reuse Rel-8 eNB receiver implementation with possible cheaper LNA
Synchronization	New item. Two solutions: (1) over LTE air interface; (2) GPS
Antennas	Lower gains compared to eNB antenna. Shorter size. Directional or omni

6.4.1 Deployment Scenarios

Due to the limited power and antenna gain of a relay node compared to a macro eNB, the number of UEs served by a typical relay node would be smaller than that of a macro eNB. This would limit the chance of using some advanced scheduling schemes to take advantage of a large number of users in the scheduling pool, for example, multi-user MIMO.

Lower height of relay antennas tends to result in richer scatterer propagation environment, where MIMO, at least rank = 2 may have more chance to be deployed. The delay spread statistics of the access link channel are not the same as of macro-UE channel, leading to different frequency selectivity. The relay scheduler should have appropriate strategies to utilize the channel characteristics specific to the access link.

Coverage extension is the most important scenario for Release 10 relay. Deploying more relay nodes to increase its density is certainly helpful to the coverage, but it may be more attractive to tweak the scheduling algorithms. Relay scheduler may sacrifice the resources to gain more coding gain, for example, to select lower modulation order and allocate more resources to distribute them in frequency. This is particular effective for the uplink transmission which is often the limiting link for coverage. TTI bundling is another way to extend the coverage, especially for VoIP traffic.

6.4.2 Relay Frame Timing

Relay frame timing depends also on the number of OFDM symbols for relay node's PDCCH. Since a relay node has to configure the subframe in its cell to be MBSFN subframe when it communicates with its donor eNB, the number of OFDM symbols for its PDCCH can be either 1 or 2 for two CRS ports and four CRS ports, respectively.

With one OFDM symbol for RN's PDCCH in MBSFN subframes, backhaul downlink transmission can start one OFDM symbol earlier, allowing one more OFDM symbol to be usable in the backhaul. Also from the deployment flexibility point of view, it is more feasible to have one or two transmit antennas at a relay node, meaning that the only the first OFDM symbol in MBSFN subframe is transmitted, with the rest being empty.

Release 10 backhaul subframe HARQ timing leads to an interesting consequence. Since Release 8 UL grant to PUSCH timing is reused in both FDD and TDD, relay node should not send a UL grant in an MBSFN subframe. Otherwise, relay served UEs would transmit PUSCH several subframes later (based on Release 8 UL grant to PUSCH timing) which collides with the backhaul uplink transmission. Obviously, a relay node should not send a DL grant in MBSFN subframes where the PDSCH region should be empty. Given the relatively small number of UEs served by a relay node, and only PHICH and DCI format 3/3a should be sent in MBSFN subframe, one OFDM symbol would be enough for L1/L2 control signaling in those subframes.

Under certain situations, for example, in order to provide the coverage over a wide rural area, it is possible to install four transmit antennas and use four CRS ports. Then, two OFDM symbols should be configured for PDCCH in MBSFN subframes.

Note that there is no signaling specified to notify the donor eNB the number of OFDM symbols in MBSFN subframes. OAM may be needed to make donor eNB aware of that information at RN.

In some relay uplink timing, the last SC-FDMA symbol in the access link cannot be received. To ensure the proper decoding of PUCCH and PUSCH, relay node scheduler should configure those access link subframes as SRS subframes, preferably cell specific SRS. Those SRS cannot be received by the relay node. Therefore, the total resources potentially for SRS are reduced, which may impact the UL frequency selective scheduling in relay cell. The RN scheduler should this into account when scheduling the uplink traffic.

6.4.3 Access Link HARQ

As discussed in Chap. 4, due to the 10 ms period for MBSFN subframe configuration, HARQ process collisions between backhaul link and access link are inevitable. So from RN point of view, once the backhaul subframes are configured, it should schedule access link transmissions, especially UL transmissions in such a way to avoid the collision with the backhaul link HARQ.

When collisions do occur, the scheduler should make the judgment on whether to prioritize the backhaul transmission or access link reception. If the decision is made to sacrifice the access link, the scheduler should carry out appropriate actions such as sending a fake NACK to trigger a retransmission.

6.4.4 Uplink Power Control for UEs in RN Cell

Relay node is a category of low power node in heterogeneous networks. Adding relay nodes to homogeneous networks (made up of macro eNBs only) would certainly change the interference scenarios. The power control parameters suitable for macro eNB-only networks may no longer be valid and a lot of optimization work is expected. Many of these work would be highly case-by-case specific, as the relay node deployment is more diversified compared to eNBs. For example, 23–37 dBm transmit power is possible, leading to quite different coverage of each relay. The flexible deployment results in different propagation environment of access link.

In LTE, an important purpose of UL power control is to regulate the inter-cell uplink interference. So certain information such as overload factor should be exchanged between macro eNBs via X2 interface. Ideally, the latency in X2 interface should be kept as low as possible, so that the uplink interference can be tightly controlled. However, the latency of X2 is generally in the order of 20–200 ms, depending on protocols and hardware processing power in the wired backhaul. For relay, the X2 interface will go through the wireless backhaul which shares the resources with access link and possibly even with macro UEs. Hence, it is expected that the X2 latency in relay backhaul would be longer than the wire backhaul. The RN scheduler should take this into account when updating the power control parameters.

6.4.5 Data Buffering

Relay is a two-hop process: downlink data originating from donor eNB are sent to a RN, then from the RN to multiple UEs; uplink data originating from multiple UEs, are sent to the RN, then from RN to the donor eNB. Data traffic are typically very dynamic, the service rates are constantly changing. However, the backhaul subframe allocation is semi-static, meaning that the time domain resource partitions do not change frequently to adapt to the traffic needs. Therefore, the data buffering in relay node scheduler is important to keep constant flow in the access link and the backhaul link, and to fully utilize the resources in each link.

Deep buffering of data would smooth out fluctuations in the traffic, and boost the spectral efficiency of access link and backhaul link. However, the delay associated with the buffering should also be carefully weighed, in particular for delay sensitive traffic such as VoIP and video streaming.

Data buffering can also change the traffic types. For example, for each UE's VoIP traffic, semi-persistent scheduling is often used, since the voice packets arrive every 20 ms with the similar packet size during a talk spurt. However, when the VoIP traffic are aggregated and with certain buffering, the data would no longer be of constant rate, since the arrival rate of a talk spurt follows a random process often modeled as Poisson. So in this sense, data buffering would affect the traffic characteristics in the backhaul.

6.5 Baseband Implementations in Donor eNB

To support Rel-10 relay functionalities, donor eNB needs to implement necessary baseband processing for R-PDCCH transmission, PDSCH transmission for RN, new DMRS generation, response to backhaul PUSCH, and new UL ACK/NACK timing.

R-PDCCH is a totally new channel. So at donor eNB, new development is needed in the areas of:

- Cross interleaved R-PDCCH.
  - To generate CCE and REG for R-PDCCH, taking into account of CRS and CSI-RS configuration.
  - To assign the interleaved DL grants to the first slots of PRBs configured for R-PDCCH transmission, and assign the interleaved UL grants to the second slots of PRBs configured for R-PDCCH transmission.
- Non cross-interleaved R-PDCCH.
  - To individually encode the DL grant and/or the UL grant for each RN, and map to the first slots and/or second slots of PRB pairs configured for R-PDCCH transmission.
  - To apply appropriate the precoder to the modulated symbols of R-PDCCH when DRMS is used for demodulation.

PDSCH module can reuse many designs of Release 8 eNB implementation. However, there are a number changes needed:

- Transmitting PDSCH only in the second slot is a new feature different from macro eNB-UE link.
- Release 10 introduces several important technologies such as DL MIMO and UL MIMO to enhance the data rate. Depending on the RN capability, donor eNB may need to implement new baseband circuit to support Release 10 MIMO features.

Rel-10 DMRS generation, especially in DL relay timing case 3 where DMRS is only present in the first slot, is a new feature and requires development. Proprietary open-loop beamforming algorithms can be implemented to take advantage of LOS propagation of backhaul and stationary RNs.

Since PHICH is not supported in backhaul link, UL grant with NDI $= 0$ should be sent when PUSCH from RN was not correctly decoded. This enables the synchronous adaptive HARQ retransmission where the resources and modulation order can be changed, compared to those in the first HARQ transmission.

In some TDD backhaul subframe configurations, the UL ACK/NACK timings are different from those of Release 8. That means in those configurations, donor eNB should expect different arrival time of UL ACK/NACK.

6.6 Scheduler at Donor eNB

Resource scheduling for backhaul data transmissions is another important module in donor eNB. In some sense, it is of more significance to the relay system performance, since a donor eNB would control multiple RNs, and perhaps co-schedule macro-UEs in the same backhaul subframe. Scheduling design requires comprehensive views of relay system designs in various aspects such as resource allocations for R-PDCCH, MCS selection, resource scheduling of PDSCH of RNs and macro UEs, etc.

6.6.1 Resource Allocations for R-PDCCH

Backhaul resources are often the bottle neck for relay system performance. R-PDCCH is considered as an overhead for the actual PDSCH transmissions for relay nodes. How to fully utilize the resources of R-PDCCH and also achieve reliable performance is an important aspect in scheduler implementation. The problem would be more complicated than the case of PDCCH. There, the detection robustness is the major goal. Since the candidate locations of PDCCH or more specifically CCEs are already determined by the Hashing function, the scheduler would only need to make the decisions on the aggregation level and which candidate space the CCEs should reside. For the aggregation level, the choices would be very limited if there is no power boosting for PDCCH. It is mainly based on CQI or other measurement report. For candidate space, the scheduler may need to arrange PDCCHs (or CCEs) of different UEs properly, to avoid excessive blockings. And this is sometimes done jointly with power boosting of PDCCH.

The above considerations for PDCCH scheduling are applicable to R-PDCCH, especially for cross-interleaved R-PDCCH where many Release 8 PDCCH designs are reused. Note that in cross-interleaved mode, R-PDCCH scheduling should be done separately in the first slots and in the second slots.

A key difference between PDCCH resources and R-PDCCH resources is that the latter can span continuously or distribute over partial or entire system bandwidth. The motivation, as discussed in Chap. 4, is to exploit frequency diversity and/or frequency selectivity, and/or precoding gain. The specification [3] only provides the vehicles, i.e., allowing various resource allocation types, and in particular, removing the burden of Hashing function for resource randomization in non cross-interleaved mode. Still, it is up to the implementation on how to achieve the goal.

There may be many ways to allocate R-PDCCH resources. Figure 6.5 just gives an example for non cross-interleaved R-PDCCH [4]. In VRB domain, total 12 resources are allocated to R-PDCCHs. For aggregation levels of 1, 2, 4, 8, there are 6, 6, 2, 2 candidate spaces, respectively. In order to take advantage of both frequency diversity and selectivity, those 12 VRBs can be distributed to a physical

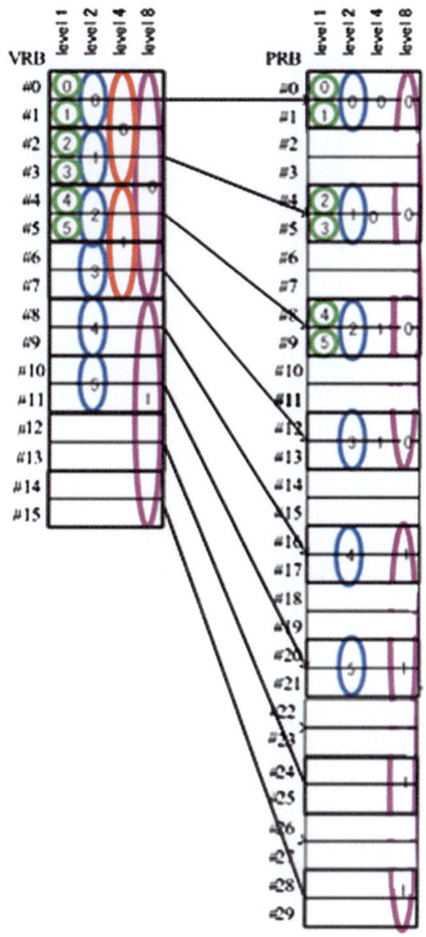

Fig. 6.5 An example of R-PDCCH resource mapping

resource region spanning over 24 PRBs, via Resource Allocation Type 1. For aggregation level 1 and 2, it is seen that their resources are localized, where frequency selective scheduling and precoding can be effectively implemented. This makes sense since low aggregation level means that the propagation environment is good and the feedback channels such as subband CQI and PMI are reliable. For aggregation level 4 and 8, their resources are distributed so that they would benefit from the frequency diversity which is important at low SNRs. Frequency selective scheduling and precoding are of less use here since the feedback channels are no longer reliable.

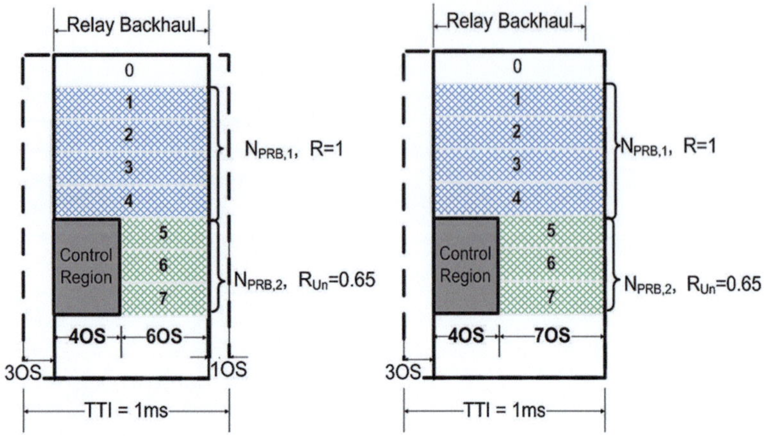

Fig. 6.6 Examples of PDSCH resource allocation patterns

6.6.2 Transport Block Size Determination and MCS Selection

Since both TDM + FDM and pure FDM are supported for R-PDCCH and PDSCH multiplexing, the physical resource patterns of PDSCH for RN are very diverse and irregular, depending on the total numbers of PRBs allocated for R-PDCCH and for PDSCH, the RGB size. The problem is further complicated by the different downlink relay timings, i.e., whether the last OFDM symbol can be received by the RN, as depicted in Fig. 6.6.

The left and right figures in Fig. 6.6 are of DL timing case 3 and case 1, respectively. The R-PDCCH carves out the gray area out of the originally regular rectangular region for PDSCH. Using Table 7.1.7.2.1-1 in [6] to get transport block size for RN's PDSCH would result in severe distortions of the code rates, since many of the PRB pairs are partially occupied by R-PDCCH.

There is precedence in TDD to solve this issue by specification. For the transport blocks transmitted in DwPTS of special subframe, since 7–10 OFDM symbols are available for PDSCH, a flat factor $R = 0.75$ is applied to adjust the difference to using regular subframes for PDSCH.

Similar procedure could be applied to R-PDCCH as proposed in [5]. It is just a matter of finer tweaking the scaling factors. Take the blue region for PDSCH in the left figure of Fig. 6.6 as an example, several factors are tried between 0.55 and 0.75 for number of PRBs ranging from 1 to 15. The empirical CDFs of code rates are compared in Fig. 6.7 [5].

The benchmark is the regular subframe that has 12 OFDM symbols for PDSCH which is shown as the blue solid curve. The purpose of this exercise is to get as much as closer to the blue curve so that the original smooth increment of code rate respect to MCS level can be maintained. Due to the partial occupancy of PRB pairs of RN's PDSCH, the code rate (as represented in transport block size) would

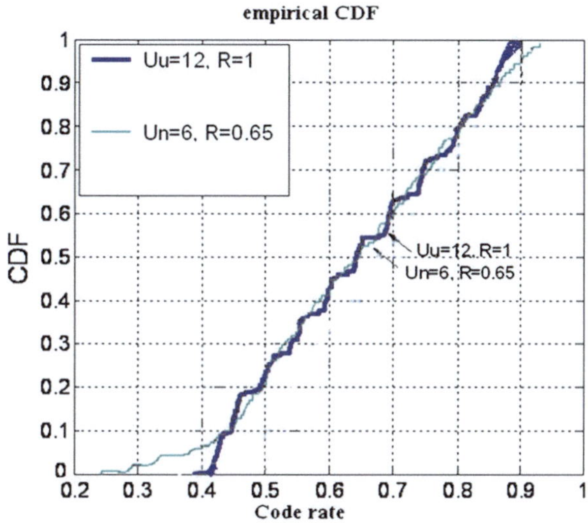

Fig. 6.7 Comparisons of CDFs of the code rate with different scaling factors to determine the transport block sizes [5]

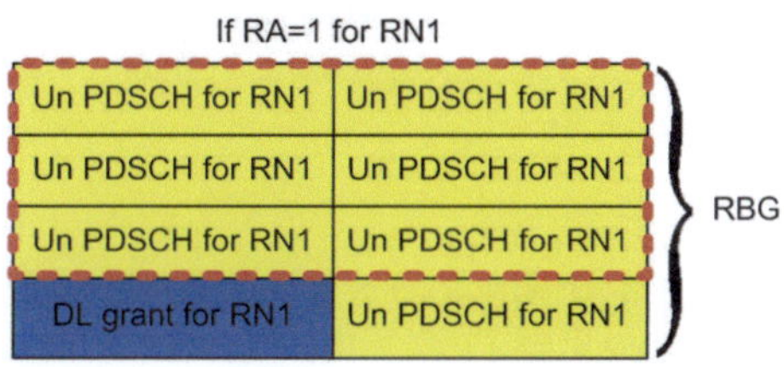

Fig. 6.8 An example of R-PDCCH and PDSCH multiplexing in a RBG (copied from Fig. 4.11a)

severely deviate from the actual MCS, if no scaling factor is to be applied. It turns out that with 6 OFDM symbols available for PDSCH in a PRB pair, the factor of 0.65 matches best the benchmark. Such factor could be specified so that different vendors would use the same mechanisms to derive the transport block sizes.

The above procedure provides a unified way to determine the transport block size. However, for RN's PDSCH, the resource patterns are so diversified that multiple scaling factors would be used to fit different situations, for example in Fig. 6.8, copied from Fig. 4.11a, where only the first slot in one PRB pair carries DL grant, while all the rest resources of total 7 slots are for PDSCH. Hence, the scheduler at eNB may need to individually calculate the number of bits that can be accommodated in the resources available for RN's PDSCH, and then check the CQI report to get MCS, and then select a closest transport block size in the table.

Take the resource allocation in Fig. 6.8 as an example. If the channel quality satisfies MCS = 18 transmission using 64-QAM, the corresponding TBS index is 16. Searching Table 6.3 assuming rank = 1 transmission, the transport block size is 1288 for regular subframes. Also assumed here are: TDD system where there are only 6 OFDM symbols in the second slot, and PDSCH starting from the fourth OFDM symbol in the first slot if that PRB pair does not carry DL grant. So the ratio of actual resources for PDSCH vs. the total resources in the RBG is roughly:

Table 6.3 Transport block size table (abridged from Table 7.1.7.2.1-1 in [6])

I_{TBS}	N_{PRB}						
	1	2	3	4	5	6	…
0	16	32	56	88	120	152	…
1	24	56	88	144	176	208	…
2	32	72	144	176	208	256	…
3	40	104	176	208	256	328	…
4	56	120	208	256	328	408	…
5	72	144	224	328	424	504	…
6	328	176	256	392	504	600	…
7	104	224	328	472	584	712	…
8	120	256	392	536	680	808	…
9	136	296	456	616	776	936	…
10	144	328	504	680	872	1032	…
11	176	376	584	776	1000	1192	…
12	208	440	680	904	1128	1352	…
13	224	488	744	1000	1256	1544	…
14	256	552	840	1128	1416	1736	…
15	280	600	904	1224	1544	1800	…
16	328	632	968	1288	1608	1928	…
…	…	…	…	…	…	…	…

$(4 \times 3 + 6 \times 4)/(10 \times 4) = 0.9$. So effectively, that RBG can carry only $1288 \times 0.9 = 1159$ bits. The closest block size in Table 6.3 is 1128 with TBS index = 14. The corresponding MCS index is 15 using 16-QAM.

Transport block size tables in [5] are based on the link level performances of AWGN channels, assuming 10 % residual block error rate after the first HARQ transmission. They do not reflect the real world link performance under various fading scenarios, HARQ strategies, interference situations. In practical systems, outer loop link adaptation is usually implemented to adjust the MCS so that it matches actual expectation of the channel rate. For backhaul channel, due to its relatively stable propagation environment and constant aggregated traffic, it is quite feasible to implement some sophisticated outer-loop mechanisms to fully handle the peculiar shapes of PDSCH resource regions for RNs (Fig. 6.6).

6.6.3 Configurations of CSI Feedback and SRS

Backhaul subframes are not continuous in time due to the half-duplex nature of Release 10 type 1 relay. Various backhaul subframe configurations (total 255 for FDD and 19 for TDD) provide flexibilities for different traffic patterns and channel conditions. Note that backhaul subframe allocations are semi-static, i.e., the scheduler cannot change them dynamically.

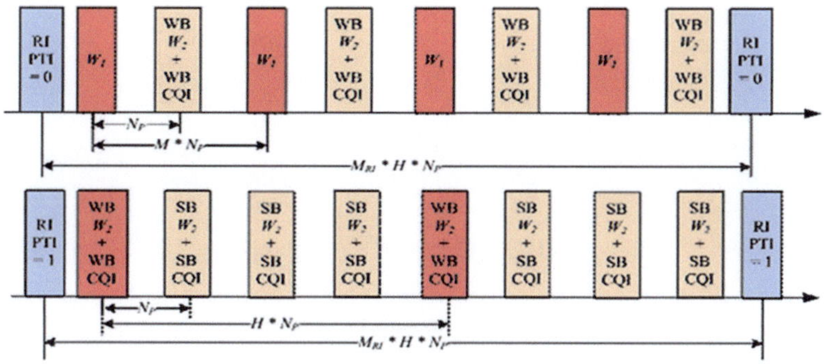

Fig. 6.9 Feedback signaling for PUCCH format 2 (3-subframe)

On the other hand, CSI feedback and SRS are often configured continuously with certain periods that do not match the backhaul configuration patterns. Some CSI feedback configurations involve the feedback transmitted in multiple uplink subframes. For example in PUCCH Mode 2 for double codebook feedback as seen in Fig. 6.9, three subframes are needed for PMI/RI feedback. CSI feedback and SRS are important for performance improvement of backhaul where the stable channel conditions encourage precoding and frequency selectivity scheduling for both downlink and uplink. CSI feedback and SRS carried in the backhaul subframes may also of macro-UEs when they are co-scheduled with RNs' data channels.

CSI feedback and SRS configurations are also semi-static. Therefore, it is not guaranteed that subframes configured for CSI report and SRS would always be uplink backhaul subframes. In the case that a subframe is not configured for RN-to-eNB transmission, the relay node should drop CSI report and SRS in that subframe.

To prevent or mitigate these situations from happening, donor eNB scheduler should carefully examine the backhaul subframe allocation patterns, and choose the appropriate period and offset for CSI feedback and SRS transmission, to strike a trade-off between the time density requirement for the feedbacks and the probabilities of the loss of the feedbacks. Both relay nodes and macro-UEs may be involved in backhaul subframes where they are co-scheduled.

Similar thing for ACK/NAKs sent in uplink subframes in TDD. With proper settings of uplink power control parameters, the uplink geometry is expected to better than that of macro UEs. Hence, it is feasible to configure ACK/NAK multiplexing on backhaul link to minimize the collisions with access subframes.

6.6.4 Resource Scheduling for PDSCH

Resource scheduling for PDSCH includes both time domain and frequency domain resources. In time domain, the resource scheduling would be rather semi-static if the backhaul subframes are used exclusively to transmit RNs' PDSCH or PUSCH.

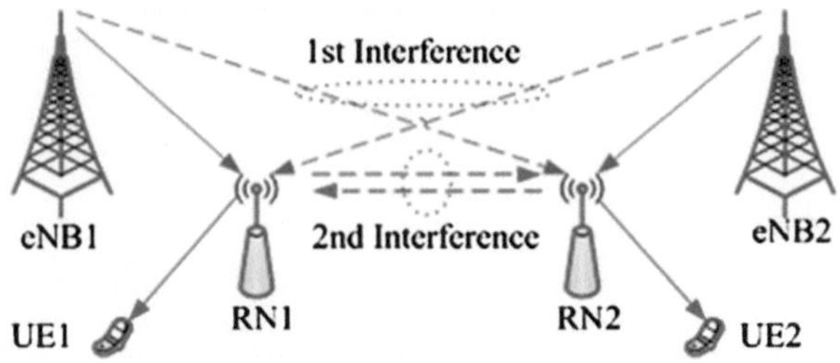

Fig. 6.10 Interference scenarios for RN at downlink backhaul

Such partition is suitable when traffic demand is stable and when the backhaul link capacity is the bottleneck of the system performance. This is a relatively clean configuration as macro-UEs PDSCH/PUSCH and the corresponding feedbacks are out of the picture.

Since the partition is semi-static, the scheduler should be careful in choosing the suitable ratios of backhaul subframes and access subframes, to fully utilize the time resources. It may not be an easy task, considering that the allocation may need to fit all RNs at least for the same donor eNB from the RN-to-RN interference point of view.

Since Release 10 relay is half-duplex operated, RN-to-RN interference can occur [7], similar to eNB-to-eNB interference in TDD, as illustrated in Fig. 6.10. If at one time, a subframe is configured as backhaul downlink subframe for RN1, and configured as the access downlink subframe for RN2, the PDSCH signals target to UE2 would cause the interference at RN1 that is receiving the data from eNB1. The interference can be severe when RN1 and RN2 are relatively close, or the propagation is LOS.

Although the time domain partition is semi-static, some resources in backhaul subframes can be used for scheduling macro-UEs, so that they can be fully utilized even for variable traffic demand. A simple way would be based on the number of UEs associated with a RN. Unless UEs are quickly moving in or out of RN cell, the information of UEs' count is rather semi-static, and can be signaled via OAM or other means. In general, the more UEs are connected to a RN, the more frequency resources are needed by that RN. Certainly, such simple partition does not consider the different access link capacities. For example, for RNs in bad propagation environment, even when it serves many UEs, the total throughput of that RN cell is still low.

Another scheme was proposed that is based on relay buffer report [8]. That report can be known to the donor eNB by certain mechanisms.

More dynamic approach would be to treat RNs the same way as macro-UEs, except that RN's data has high priorities. Similar proportional fairness scheduling can be carried out and delay sensitivity metric can also be added.

6.6.5 *Open Loop Uplink Power Control for RNs*

Although the uplink maximum transmit power of RN is comparable to that of UE, e.g., 23 dBm, the actual needed power would be much lower, considering the relatively good channel condition of the backhaul and directional antennas at RN uplink transmitter. Even with low transmit power, the interference caused by RN's uplink transmission to neighboring eNBs could be significant when the channels connecting to those eNBs are also good or LOS dominant, and within the radiation direction of RN uplink transmit antennas. All these are quite different from regular UEs that have less touch with either its serving eNB or the neighboring eNBs. Uplink power control parameters should be modified compared to those for macro UEs.

Given the relatively wider coverage of uplink signals from RN, interference over thermal (IoT) at donor eNB tends to have more fluctuations in time and frequency. Hence, more frequent information exchange on the overload may be needed via X2 interface. This could require more effort if PUSCH of RNs and macro-UEs are co-scheduled in backhaul uplink subframes.

6.7 Relay Network Planning

Relay nodes would be deployed by wireless service operators. Basic ideas resemble the cell planning for macro eNB deployment. However, there are a few RN specific characteristics we need to pay attention.

6.7.1 *Number of RNs*

As a module of network equipment, relay planning shares a lot of similarities to macro eNB planning. Release 10 relay is mainly for coverage extension, therefore, relay network planning can target for coverage, with throughput optimization as the second priority. UE outage is a good measure of coverage and can be used for relay planning. In [9] an example is provided for the coverage analysis of a real network. Although the pathloss and shadow fading are based on 3GPP/ITU models described in Chap. 2, the study would capture the essential behavior of the relay network.

The total area is 6.8 km long (from west to east) and 4.1 km wide (from north to south) where 108 macro cells (about 36 eNB sites) are deployed. The relay is deployed in 2.6 GHz. The maximum transmit power of RN is 37 dBm. When there are only macro eNBs in the network, the outage probability on average is about 11.2 %. When 40 relay nodes are deployed in the network, the overall outage rate is reduced to 7.7 %.

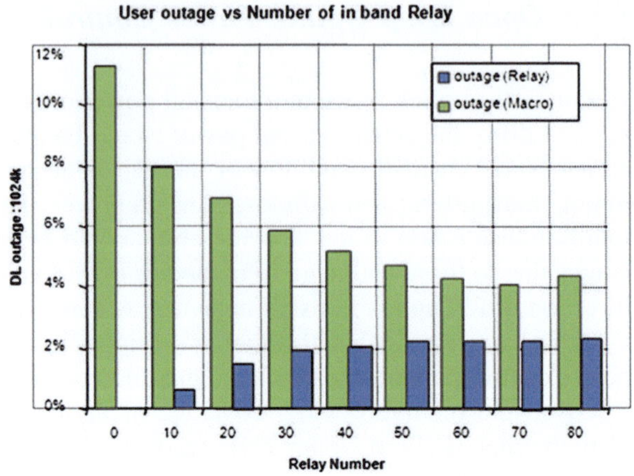

Fig. 6.11 Outage levels with different numbers of relay nodes

Figure 6.11 compares the outage probabilities of relay cells and macro cells when different numbers of relay nodes are deployed. It is observed that as the number of relay nodes increases, the macro cell outage decreases significantly, since most of those UEs in outage are switched to relay nodes. However, the relay cell outage also increases, at a slower rate. This may be contributed by the limited coverage of each RN, and limited backhaul capacity.

The lower transmit power and antenna gain result in smaller coverage of a RN compared to an eNB. Hence, in certain areas, adding eNB or adding another carrier to the original eNB may be more attractive than adding a number of relay nodes, if such measure is cost effective. Therefore, the overall planning should consider the cost structures of macro eNBs and relay nodes, so that minimum cost is needed to ensure certain target outage rate.

6.7.2 RN-to-RN Interference

Another important aspect of relay network deployment is relay node synchronization and backhaul subframe allocation. Poor synchronization or wrong configurations of backhaul subframe would cause significant interference. Such lessons were already learned during the deployment of TDD systems.

In FDD systems, the network can be operated in either asynchronous or quasi-synchronous mode. This may not be a problem in macro cell only systems, although quasi-synchronous would benefit some processes such as cell searching and handover. However, for the half-duplex relay, the interference issue arises for example in the scenario illustrated in Fig. 6.10.

Let us consider 3GPP Case 1 whose inter-site distance is 500 m and Tx power of RN is 30 dBm. DL transmit antenna at RN for access link is omni-directional in horizontal. RNs are randomly placed in a macro cell, without site optimization.

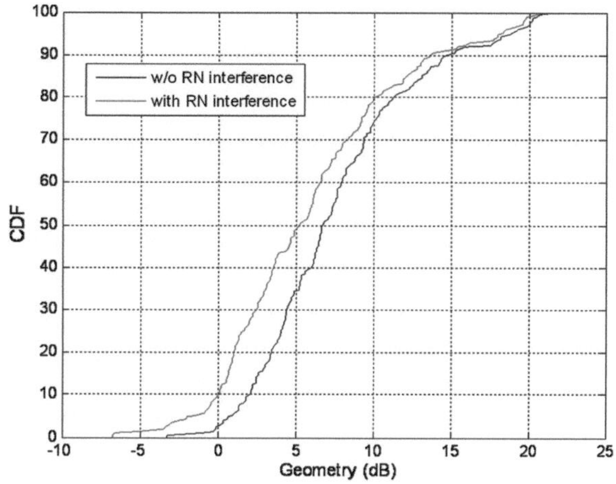

Fig. 6.12 Backhaul DL SINR with and without RN–RN interference in quasi-synchronous FDD systems

RN–RN pathloss model reuses RN-UE pathloss model. The same backhaul subframe allocation is assumed for all RNs in a macro cell, therefore, no RN–RN interference comes from RNs of the same cell, if we ignore the small differences in propagation delays of the backhauls for different RNs. That timing difference would just cause the interference of a fractional of OFDM symbol, considering the rather small size of the macro.

Quasi-synchronous FDD operation is considered where the synchronization error is about one OFDM symbol of duration (~70 μs). Hence 10 % of the subframe is overlapped or interfered. The simulation result in Fig. 6.12 shows that in LOS propagation environment, the interference can significant degrade the SINR at RN.

Note that the result may be a little optimistic since RN-UE model is reused for RN–RN pathloss. In reality, there would be more LOS and less attenuation in that path. Also note that such interference may not be effectively reduced by RN antenna down-tilt. Since the Tx antenna gain of RN access link is only 5 dBi, its vertical radiation beamwidth is expected to be fat, which is less sensitive to the down tilt.

Above analysis seems to indicate that tighter synchronization, i.e., ~70 μs, between eNBs is needed in FDD to avoid significant RN–RN interference between cells. The result also clearly suggests using the same backhaul subframe allocation for RNs in all macro cells that support relay operation.

6.7.3 Cell Range Expansion and ABS Configuration

Similar to pico or femto node, Release 10 relay would benefit from cell range expansion and ABS configuration. In [10], simulation study was conducted to

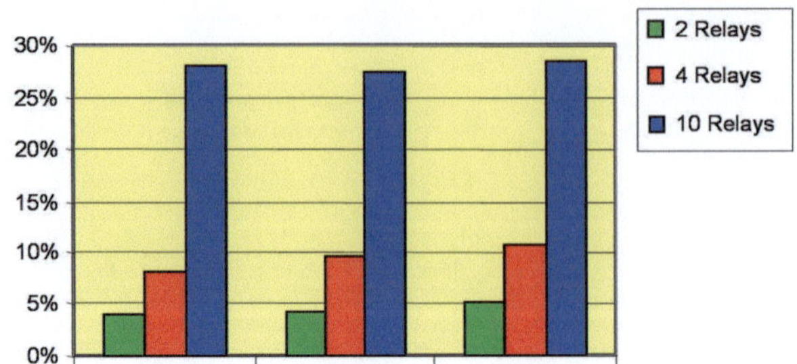

	5% UE	50% UE	Mean
2 Relays	4%	4%	5%
4 Relays	8%	10%	11%
10 Relays	28%	27%	29%

Fig. 6.13 Relay performance gains in Config. #4 with range expansion

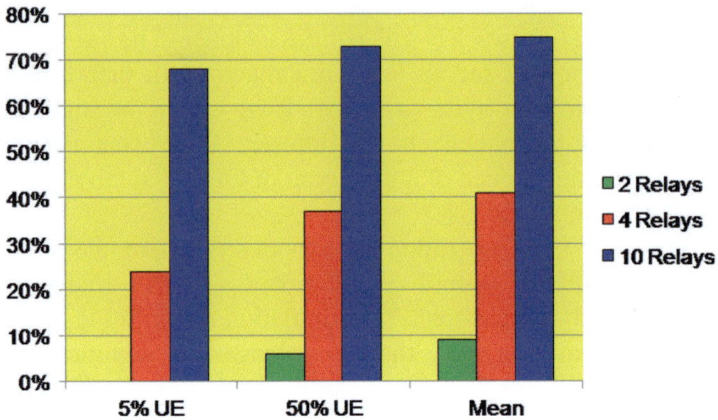

Fig. 6.14 Relay performance gain in Config. #4 without range expansion

evaluate performance gains of relay under various conditions. As Fig. 6.13 shows, compared to the gains in Fig. 6.14, the range expansion can further improve the average system throughput by about 4, 30 and 46 % for 2, 4, and 10 relays per cell, respectively.

Similarly, almost blank subframe (ABS) can be configured in relay deployment when large bias is applied. ABS is more configured in macro cell since macro eNB transmission would cause more interference than low power node. There is more freedom to configure ABS in the case of pico or femto node deployment, since their backhaul does not consume wireless resources. However, for Release 10 relay, there is a chance that ABS overlaps with backhaul subframe. Since there is no procedure at the relay node to handle the potential collisions, eNB scheduler should try to avoid that situation from happening, by imposing certain constraint on subframe allocations either for backhaul or ABS.

It should be pointed out that due to the limited capacity of relay backhaul link, the gains from range expansion and ABS would be significantly less compared to pico node or femto node. These enhancement features should be carefully examined by taking into account the backhaul setting, such as number of antennas, relay site optimization.

References

1. 3GPP TR 36.814: Evolved Universal Terrestrial Radio Access (E-UTRA): Further advancements for E-UTRA physical layer aspects
2. 3GPP R1-100559: Further consideration on relay channel modeling. RAN1 #59bis, CMCC, Jan 2010
3. 3GPP TS 36.216: Evolved Universal Terrestrial Radio Access (E-UTRA): Physical layer for relaying operation
4. 3GPP R1-105176: Design of non-interleaving R-PDCCH in Rel-10. RAN1 #62bis, CATT, Oct 2010
5. 3GPP R1-106316: Un TBS determination. RAN1 #63, CMCC, CATT, Nov 2010
6. 3GPPTS 36.213: Evolved Universal Terrestrial Radio Access (E-UTRA): Physical layer procedures
7. 3GPP, R1-101384: Considerations on backhaul interference and synchronization for relay. RAN1 #60, CMCC, Feb 2010
8. 3GPP, R1-106225: On the performance evaluation and improvement for relays. RAN1 #63, NEC Group, Nov 2010
9. CMRI-NSN: Relay cost analysis performance and cost evaluation of relay deployment in CMCC Beijing networks. Nokia-Siemens Networks, CMCC, Dec 2010
10. 3GPP R1-094893: Updated type 1 relay performance characterization: Dependency with channel model assumptions. RAN1 #56bis, Qualcomm, Mar 2009

Chapter 7
Outlook of Relay in Future LTE Releases

7.1 Some Trends in Mobile Communications

7.1.1 Trends at Terminal Side

Cell phones, as the most common form of terminal for mobile communications, become more powerful in various aspects:

- Carrying more smart applications and vastly increasing the usefulness and functionalities of the terminals, well beyond for voice communications and short messages.
- Ever increasing processing capabilities with the continuing size shrinking of integrated circuits. A smart phone is like a personal computer.
- Mobile social networking and mobile advertising. Proximate services to discover friends in the vicinity and find people that share common interests. Device-to-device (D2D) communications is one example [1, 2]. As Fig. 7.1 shows, the base stations may or may not directly participate in data transfer between users in D2D, such as allocating uplink resources for originating UEs and reception, and allocating downlink resources for target UEs and transmission. Instead, the network would just do some control over D2D communications. Such control can be very loose and at very high level, or it can be very tight, down to L1 level. Nevertheless, since the traffics do not go through the network, security is a major concern for D2D.

7.1.1.1 Trends at Network Side

From the network side, we witness the migration from pure macro-eNB based homogeneous networks to macro and low power node combined heterogeneous networks. The capacity improvement by service operators cannot keep up with the

Y. Yuan, *LTE-Advanced Relay Technology and Standardization*, Signals and Communication Technology, DOI: 10.1007/978-3-642-29676-5_7,

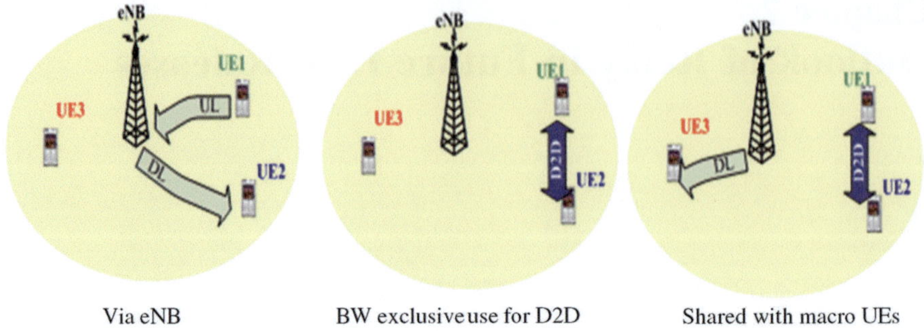

Fig. 7.1 Device-to-device (D2D) communications

explosive growth of the traffic. Therefore, offloading traffic to low power nodes such as pico node, femto node, or even to the terminals becomes more attractive. There are two major approaches on how to use low power node for capacity enhancement:

- Cell splitting

Deploying more pico node, femto node or Release 10 relay node should help to achieve the cell splitting gain. The gain can be further improved by cell range expansion. However, it comes with the price of strong interference, especially from macro transmissions, since these nodes have their own resource scheduler usually independently running. Time domain resource coordination such as configuring almost blank subframe (ABS) can mitigate the interference. Still, basic signals such as common reference signals, paging and synchronization channels, primary broadcast channels are not protected by ABS. Ultimately, cell splitting approach requires strong interference cancellation capabilities at the terminal.

Since mobile processing power is ever growing, more advanced signal processing techniques are becoming feasible, which allows UEs operating in severe interference environment due to the cell splitting.

- Inter-node cooperation

The general inter-node cooperation is CoMP. Here, we specifically refer to CoMP Scenario 4 where the low power node, typically remote radio head (RRH), shares the same cell ID with the macro eNB. Joint transmission and reception can be carried out at multiple nodes, i.e., macro and RRH. Since the resource scheduling is centralized, the number of participating nodes for joint transmission/reception can change dynamically, to better adapt to the traffic variations. Same cell ID RRH appears transparent to UEs, thus the cooperation between macro and RRH constitutes a virtual macro cell of distributed antennas whose coverage shape can constantly change, i.e., “soft cell” [3].

To fully achieve "soft cell", the traditional common reference signal (CRS) based radio resource management (RRM) needs to be changed. CRS is cell specific and common to all UEs belonging to the same cell. However, the cooperation between macro and RRH is UE specific. In another word, the virtual soft cell is UE specific. In this respect, CSI-RS can be used for RRM, if it is configured as UE specific. This is a fundamental change not only at physical layer specification, but also at higher layers, since RRM affects how UE's mobility is handled, which involves a lot of higher layer signaling and procedures during the handover.

Removing the reliance on CRS for RRM means that Release 8 PDCCH would no longer be used for L1 control signaling which is based on CRS. The enhanced PDCCH (ePDCCH), currently in the process of standardization [4], may serve the purpose. Frequency domain multiplexing nature with demodulation reference signal (DMRS) allows more flexible resource allocations and increased capacity for L1 control signaling.

- "Cloud" RAN

The base station in traditional wireless network is essentially a piece of standalone equipment with all the necessary baseband capabilities and RF functionalities. The cloud-RAN concept is changing this traditional setting, and advocating centralized baseband processing, an analogy to cloud computing. Its effect is far-reaching, not only on the business model of operators and product plans of telecommunications equipment vendors, but also technology evolutions in future mobile communications.

Cloud RAN is sometimes dubbed as CRAN to emphasize its centralized, clean, cooperative, cloud based nature. CRAN features centralized baseband in a big processing pool. Local baseband processors become unnecessary, therefore saving the expensive air conditioning to maintain the normal operations of the baseband. The air conditioning cost contributes the most percentage in power usage of a base station.

Centralized baseband can serve as a platform supporting multiple radio access technologies. The platform is open for the access since the processing is performed on general purpose servers. Through software (re)configuration and upgrade, different technologies can be easily added in, including the future specifications. This is very beneficial to technology evolutions as equipment vendors and operators do not need to worry about out-date of their hardware investment in previous technologies.

Figure 7.2 shows the network elements in CRAN. The baseband processing is carried out in virtual base station clusters which consist of general-purpose processors to perform PHY/MAC processing. The inter-cluster communication is through X2+ interface. The high speed switching can dynamically balance the traffic load of the network, to maximize the computation efficiency between the clusters, and between the centralized cloud and radio. In the field, a large number cooperative remote radio units (RRUs) can reduce interference and achieve high spectral efficiency.

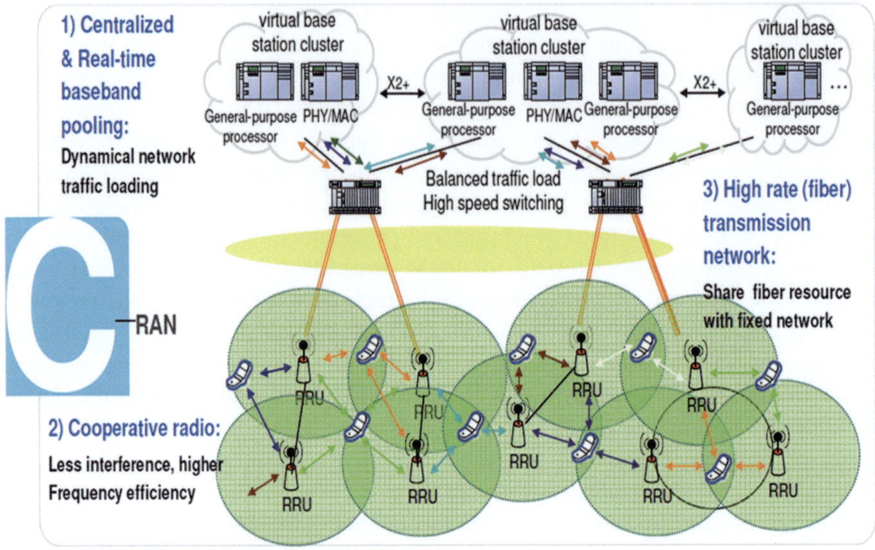

Fig. 7.2 Network elements in CRAN

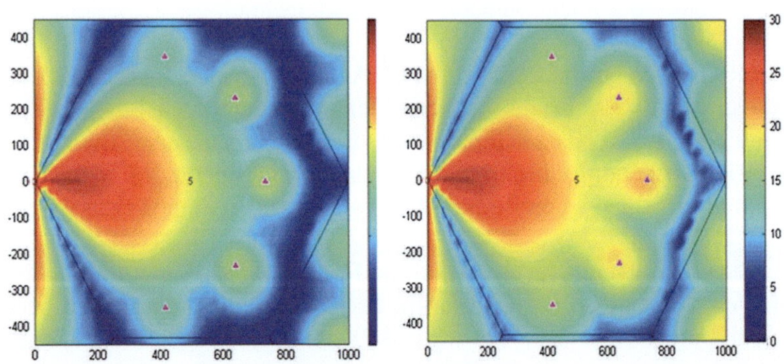

Fig. 7.3 Change of virtual cell shape with cooperation

7.2 Cooperative Relays

Relay node can be cooperative. Type 2 relay studied in Release 10 is one example. Sharing the same cell ID with macro node makes the operation of type 2 relay analogous to the same cell ID remote radio head (RRH), with the only difference in backhaul, fiber optic vs. wireless. While from system capacity prospective, cooperative relay cannot compete with fiber connected RRHs, wireless backhaul allows much more flexibility of the relay deployment, not only with fixed locations, but also with nomadic movement or completely mobile. Cooperative relay

Fig. 7.4 Network coding applied in cooperative transmission

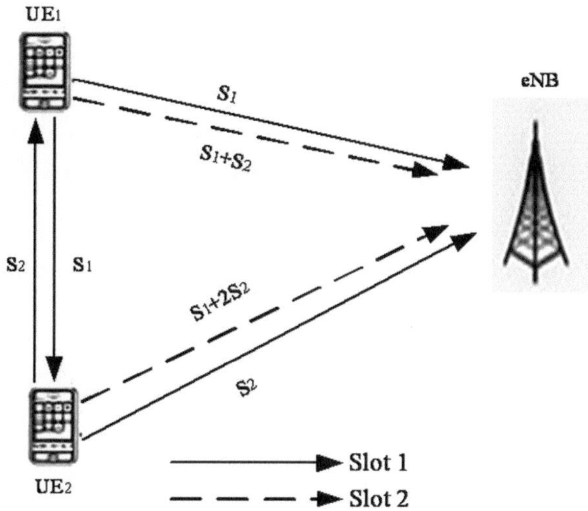

node and macro eNB dynamically form a virtual cell whose shape can change like fluid or amorphous material, as seen Fig. 7.3.

Previous study on type 2 relay was constrained by the backward compatibility for Release 8 UEs, thus closing the door for more advanced features potentially helpful for the performance. For example, type 2 relay does transmit Rel-8 CRS, leading to the pessimistic CQI estimation for combined channel in the case of cooperative transmission, or the totally different CQI estimation for resource reuse. Such mismatch can only be handled by eNB implementation, i.e., outer loop link adaptation. Release 8 HARQ timing in backhaul prevents some more efficient mechanism for cooperative relay in the uplink.

Such backward compatibility is no longer the limiting factor. With the introduction of UE specific CSI-RS and enhanced PDCCH, there are more freedoms for design optimization of cooperative relay. From this respect, some on-going work in LTE Release 11 of enhanced PDCCH, UE specific CSI-RS and power control for uplink CoMP could be reused for cooperative relay to improve its performance.

Cooperative relay is not limited to those already been studied in Release 10. More widely use of network coding is a promising direction. In the context of network coding, the cooperative relaying operation can also involve UEs as Fig. 7.4 shows, as long as they can participate in relaying. In this sense, cooperative relay can also be used in D2D communications. In Fig. 7.4, in addition to transmitting its own data to eNB in the first slot, UE1 and UE2 can try to decode each other's data during the first slot and pass them to eNB in the second slot. Through this cooperation, significant gain is observed in Fig. 7.5.

Network coding not only brings capacity gains, but also improves the multipath diversity and energy efficiency. Certainly, there are quite a few challenges in applying network coding to cooperative relays. For example, strict synchronization is required among sources participate in the cooperation. The performance is

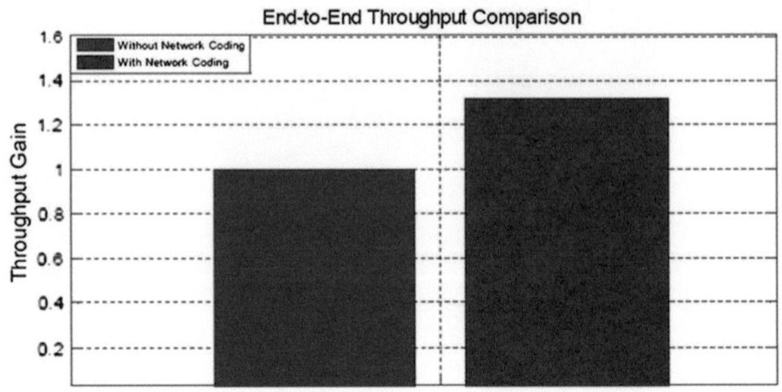

Fig. 7.5 Uplink cell throughput gain with network coding based cooperative transmission

Table 7.1 Achievable spectral efficiency of the cooperative relay compare to the capacity

Bits/Symbol SNR(dB)	Capacity of cooperative relay	Cooperative relay rate achieved via LDPC	Non-cooperative capacity	Rate achieved of non cooperative relay
−20	0.04	–	0.02	–
−12.5	0.14	0.1	0.05	–
−10	0.23	0.18	0.07	–
−9	0.26	0.2	0.09	–
−7	0.3	0.3	0.11	–
−5	0.45	0.4	0.2	–
−4	0.5	0.44	0.22	0.2
−3	0.59	0.5	0.24	0.25
−0.9	0.66	–	0.42	0.38
0	0.68	–	0.5	0.44
2	0.76	–	0.58	0.6
5	0.9	–	0.96	0.95
8	0.97	–	0.96	0.95
10	1	–	1	–

highly relying on source grouping methods which should be optimized, yet also efficient. The control signaling overhead should be carefully considered so that it would not eat out the potential gains in data transmissions.

Network coding based cooperative relay also opens the door for new channel coding. Besides the legacy Turbo codes, LDPC codes prove to be a good candidate as it has more flexibility to adapt to different scenarios of operations. An example is shown in [5] where the rate-compatible LDPC codes have been optimized for the two-hop cooperative relaying. The codes are irregular and designed based on edge growth and parity splitting. For Table 7.1, it is seen that the performance of the LDCP codes is quite close to the capacity of the cooperative relaying.

Any channel coding would be an overhaul from physical layer specification prospective. Hence, extensive study is needed for any newly proposed coding scheme.

7.3 Relay Backhaul for High Speed Mobility

During the study phase of LTE-Advanced, group mobility was identified as one of the key application scenarios for relay deployment. Relay node is more suitable for group mobility due to the following:

- Compared to repeater

Relay node in general is of decode-and-forward type, thus can improve signal to interference and noise ratio (SINR). Compared to a repeater that equally amplifies both signal power and interference power, a relay node allows separate optimization of backhaul link (donor eNB to relay node) and access link (relay node and the UE), and has the potential of improving the link capacity.

Repeater requires much less standards work compared to relay node, especially in RAN1. However, the link capacity issue may significantly limit repeater's use in group mobility scenario where the target is more on capacity enhancement, rather than to overcome the excessive thermal noise.

- Compared to regular UE

Relay node is usually not powered by battery and has less constraint on transmit power compared to regular UE. More advanced and power-consuming baseband processing is affordable in relay node. More antennas can be mounted on a relay node than a regular UE, especially if the UE is a hand-held device. Given the less limitation on its size, directional antenna (both vertical and azimuth) is possible for each antenna element on a relay node, whereas the regular UE antennas are omni-directional in azimuth and very fat in vertical.

All above differences from regular UE give relay node much more potential for spectral efficiency improvement, which is important for group mobility scenario.

Passengers on high-speed train are more likely to be data-hunger professionals and would hook up to the internet & emails when on-board (voice call is considered impolite here). The capacity requirement for relay backhaul is expected to be very high, when the very high user density is considered in a train. The user density would be high even when it is not fully loaded. The high data rate expectation is applicable for both downlink and uplink traffic.

Backhaul channel characteristics, including pathloss, shadow fading and fast fading, would be different from those of eNB to UE connection, due to

- Higher elevation of mobile relay antennas mounted typically on top of train roof ($\sim$5 m)
- Terrain and morphology along the rail track has less scatterers

As discussed above, increasing the backhaul channel capacity should be the main concern of mobile relay for group mobility, especially on fast-moving vehicles.

- Given the less power constraint on relay node, wider bandwidth can be considered in the backhaul with more flexible solutions for carrier aggregation.

- Multi-antenna technologies can be further optimized to fit the mobile relay node capabilities and the propagation environment along the track lines
- Control and signaling channel optimization to improve the reliability and link robustness

The enhancement over backhaul will have no impact on access link to UE. Standard can be kept untouched for UE side as we done in Rel-10 relay. The legacy LTE handset can well access the network without awareness of mobile relay.

7.4 Cooperative Mobile Relay

Device-to-device communication can be considered as a simple mobile relay—a moving terminal disseminating data to nearby terminals. Mobile relay provide more "bridges" to more efficiently transfer the traffic between terminals, and between terminal and eNBs/pico/HeNBs. Mobile relay is often called mesh ad-hoc wireless network that captures a lot of attention from the academia. It is also found use in military applications where centralized networks are generally not available, or too insecure. Even though there is still a long way before the technologies in ad-hoc wireless network would be practical enough to be considered in wireless industry, it reflects the future trend of mobile communications.

Type 2 relay studied during Release 10 is a fixed relay. Unlike type 1 relay which has its own cell id and is more difficult to handover between donor eNBs, type 2 relay has no cell id. This makes it easier to handover between neighboring eNBs. Moving relays have the advantage of achieving more seamless coverage and capacity enhancement. One particular application would be bus-mounted cooperative relay as illustrated in Fig. 7.6. Not only to serve the passengers on the bus, the mobile cooperative relay can also assist data communications for nearby users outside the bus, i.e., pedestrians on the sidewalk. Buses are usually powered by the gas engines or power grid, making transmit power of mobile relay node less an issue. The routes of buses are with high density of populations and type 2 relay can help to boost the network capacity. The traveling velocity of a city bus should be slow to medium, i.e., <40 km/h, so the Doppler is not expected high as fast speed trains.

7.5 Local Server

As discussed earlier in this chapter, the difference between relay and UE starts to be blurred in future mobile communications. Powerful UEs will be able to perform relay functionalities, while relay node can be nomadically deployed, or even with mobility. Device-to-device (D2D) communications, while promising for proximity services, has its own drawback, for example:

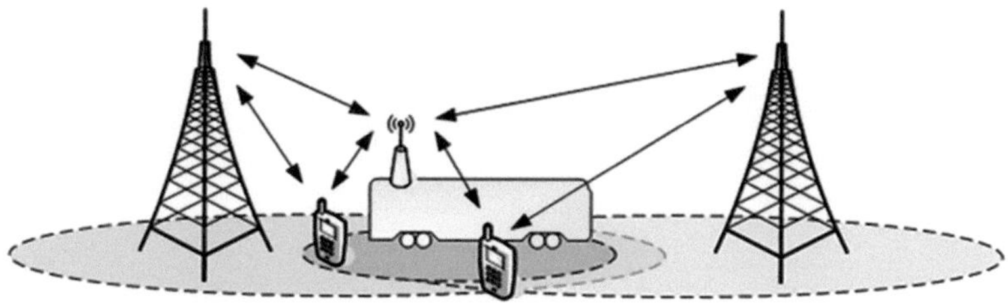

Fig. 7.6 Bus-roof mounted mobile cooperative relay node

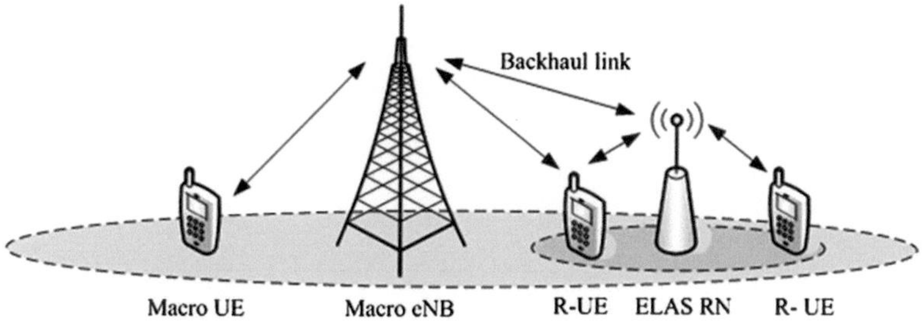

Fig. 7.7 An example of relay based wireless local server

- High requirement for terminals, significant changes at physical layer and upper layers
- Difficult to monitor the information exchanged between D2D users. Big concerns of security
- Size of terminals limits transmission rate, the power consumption and the coverage of users engaged in D2D communications

Alternatively, a wireless local server can be deployed within the coverage of macro cell to enhance local content based services, as shown in Fig. 7.7.

Such server can be based on relay, either type 1 relay or cooperative relay. It can perform data relaying between users, or multicast data to local users of the same group. The local services include advertisement, public information broadcasting in supermarkets, restaurants, hospitals, local media downloading, user data storage/sharing, wireless payment, food ordering, wireless multimedia tour guiding, push-to talk, group mobile gaming, etc.

References

1. Wireless World Initiative New Radio-WINNER+, Innovative concepts in peer-to-peer and network coding
2. Wu, X. et al.,: FlashLinQ—a synchronous distributed scheduler for peer-to-peer ad-hoc networks. 48th Annual Allerton Conference, pp. 514–521. (2010)

3. Soft cell, a HetNet solution for IMT-advanced and beyond. Ericsson Research, Nov 2011
4. 3GPP RP-111776: Enhanced downlink control channel(s) for LTE, RAN #54, Alcatel-Lucent, Alcatel Shanghai Bell, Dec 2011
5. 3GPP R1-083852: TDMA relay code for half-duplex L2-based relay in LTE-Advanced, RAN1 #54bis, Alcatel-Lucent, Sept 2008

MIX
Papier aus verantwortungsvollen Quellen
Paper from responsible sources
FSC® C105338
FSC
www.fsc.org

If you have any concerns about our products,
you can contact us on
ProductSafety@springernature.com

In case Publisher is established outside the EU,
the EU authorized representative is:
Springer Nature Customer Service Center GmbH
Europaplatz 3, 69115 Heidelberg, Germany

Printed by Libri Plureos GmbH
in Hamburg, Germany